1 9 9 6
YEARBOOK

Communication in Mathematics, K–12 and Beyond

Portia C. Elliott

1996 Yearbook Editor
University of Massachusetts at Amherst

Margaret J. Kenney

General Yearbook Editor
Boston College

NATIONAL COUNCIL OF TEACHERS OF MATHEMATICS

Copyright © 1996 by
THE NATIONAL COUNCIL OF TEACHERS OF MATHEMATICS, INC.
1906 Association Drive, Reston, Virginia 22091-1593
All rights reserved

Library of Congress Cataloging-in-Publication Data:

Communication in mathematics, K–12 and beyond / Portia C Elliott,
1996 yearbook editor, Margaret J. Kenney, general yearbook editor.
     p.   cm. — (Yearbook ; 1996)
   Includes bibliographical references and index.
   ISBN 0-87353-423-9 (hc)
   1. Mathematics—Study and teaching.   2. Communication in
education.   I. Elliott, Portia C.   II. Kenney, Margaret J.
III. Series: Yearbook (National Council of Teachers of Mathematics)
; 1996.
QA1.N3   1996
[QA11]
510′.71—dc20   96-2675
                                CIP

Printed in the United States of America

Contents

Preface

As we move rapidly into the next millennium, with information technology growing exponentially, it will not be enough to know a few isolated facts in arithmetic. Children will need to make sense of, and communicate in, the quantitative world they are inheriting. This mathematics sense-making will require them to read, write, experience, explain, discuss, defend, and clarify for themselves and others what are sure to be cognitive dissonances in their realities. The teaching of reading, writing, and arithmetic in this increasingly symbol-laden society will need to be shifted to reading and writing *about* mathematics, if true mathematics literacy is to be achieved.

This shift in content emphasis, with a concomitant shift in classroom discourse, will raise questions in the minds of those who believe that textbooks should control the flow of ideas in the classroom, not student-defended conjectures and experience-gathered evidence. This shifting will also be problematic for those who want the standardization of curriculum and assessment to facilitate quantifying and seriating achievement, not standards-based teaching and assessing that call for valuing all "points of view" in the mathematics meaning-making process. Critics and supporters alike will need to justify these shifts using the best arguments we can articulate. They will need forums to express their concerns and assure themselves that children are learning the mathematics that will enable them to communicate in the "global market place."

Who best can help move a nation and its children to "spreadsheet" thinking and mathematics meaning-making? The Advisory Panel of this yearbook believe the answer is communities of people (students, teachers, and others) willing to have open, honest, constructive dialogues about mathematics and what it means to be mathematically literate. As children begin to experience the power of their mathematics literacy, they will question an educational system that gives power and privilege to a few while relegating an inordinate number of people to lives of mathematical inarticulateness. Parents and others will demand to know in unambiguous terms, "Is anybody out there learning mathematics?" Helping both groups use their power of mathematical thinking and communicating should help them determine the truth or fallaciousness of arguments. After all, this is the essence of mathematics communication.

Because the classroom teacher is pivotal in creating, explaining, and championing shifts in content and discourse, the heart of this book contains ideas for teachers facing the challenges of turning their classrooms and schools into "discourse communities." The yearbook is divided into four sections. Part 1 (chapters 1–3) sets the stage by considering the challenges inherent in shifting directions of discourse. Part 2 (chapters 4–21), the heart of the book, focuses on establishing discourse communities within the classroom. Subsections of Part 2 are devoted to examining modes and tools of discourse and ways to assess students' efforts at discourse. Part 3 (chapters 22–25) moves the discourse discussion outside the K–12 arena by considering the importance of discourse in wider audiences, namely, schools of education, homes, and community groups. Finally, Part 4 (chapters 26–28) focuses on the role of language in mathematics discourse. Authors in this section argue for allowing children to learn the language of mathematics as naturally as they learn their native language.

The production of this yearbook was made possible by the tireless contributions of numerous people. My sincerest appreciation goes to my superb Advisory Panel:

Frances Curcio	New York University, New York, New York
Marvin Doubet	Lake Forest High School, Lake Forest, Illinois
Eric Hart	Western Michigan University, Kalamazoo, Michigan
Sheila Sconiers	University of Chicago, Chicago, Illinois
Margaret Kenney *General Yearbook Editor*	Boston College Mathematics Institute, Chestnut Hill, Massachusetts

Their knowledge and professionalism were eclipsed only by their sensitivity to, and appreciation for, the enormity of the challenge of moving a nation to mathematical literacy. I also gratefully acknowledge the assistance of Cynthia Rosso, Charles Clements, and the production staff at the NCTM Headquarters Office in Reston for their outstanding editorial and production contributions. Finally, I would like to express a special thanks to the authors who responded to our first call for manuscripts and to the fifty-five authors (twenty-eight manuscripts) that make up this volume. It is the effort of all these dedicated professionals that have put this book into your hands as a resource for thinking about "communication in mathematics, K–12 and beyond."

Portia C. Elliott
1996 Yearbook Editor

1

Communication— an Imperative for Change: A Conversation with Mary Lindquist

Mary M. Lindquist

Portia C. Elliott

In THIS conversation with Mary Lindquist, president of the National Council of Teachers of Mathematics 1992–1994, and Portia Elliott, editor of this yearbook, Lindquist shares some of her insights into the role communication plays in mathematics education reform. Readers are reminded of the challenges we in the mathematics education community face in attempting to communicate what we value, believe, and know about helping a nation and its children become mathematically literate. We have chosen this conversation format because it lends itself to our situation— two colleagues communicating (by fax and e-mail) about teaching, assessing, and learning mathematics in a yearbook on communication. What could be more appropriate?

Mary, in your tenure as president of NCTM, you have traveled the country (probably more miles than you care to calculate) facilitating conversations among various stakeholders on many aspects of school mathematics reform. It is apparent that you feel dialogue is important. Could you please explain why you say communication is "an imperative for change"?

First and foremost, Portia, if we want to fulfill the societal goals of a mathematically literate workforce, lifelong learning, opportunities for all, and an informed electorate, then we all will need to communicate mathematically. These goals are so fundamental to the work of the Council that learning to communicate mathematically is one of the five overarching goals for students.

There is another reason why communication is so important. Channels of communication must remain open and bidirectional so that the Council's vision can be articulated and concerns about proposed changes can be raised and, we hope, resolved. Only in sharing our efforts, successes, and struggles—and hearing from both critics and constituents—can we hope to elicit the support of the many communities that will need to embrace school mathematics reform to make it a reality. Discussion will be our greatest ally as we proceed.

Many involved in the reform movement have written about the shifts in communication taking place in mathematics classrooms. In this yearbook, we have fifty-five authors who lend their voices to this discussion. Most teachers would probably say, "Sure, my students and I communicate all the time. So what's the fuss?" Mary, share with us the Council's reasons for making communication an explicit part of each of its three Standards *documents.*

I believe that it is evident why communication played such a central role in the NCTM's *Professional Teaching Standards,* since teaching is communicating. It is *how* we communicate that is given direction by the teaching standards. It reminds me of the saying by Ralph Waldo Emerson that "conversation is the laboratory and workshop of the student."

A quote from the *Assessment Standards* clearly sets the centrality of communication to assessment: "Assessment is a communication process in which assessors—whether students themselves, teachers, or others—learn something about what students know and can do and in which students learn something about what assessors value" (NCTM 1995, p. 13). Throughout the document are examples of how we communicate what we know about students' learning to a variety of audiences. It caused me to think about the connection between the purpose of the assessment and what and how we communicate.

If we consider that mathematics is a language and that this language is best learned in a community of other learners, then it is easy to understand why there is a communication standard in each of the three levels of the Curriculum Standards and a matching standard in the Evaluation Standards.

To answer your question more directly, communication is explicit in the three *Standards* documents because it is the essence of teaching, assessing, and learning mathematics.

It is clear why you believe communication must be an essential element both "inside" and "outside" classrooms if school mathematics reform is to be successful. Let's talk for a minute about communication in the classroom. One of the truisms handed down in my family from my grandfather, a college mathematics teacher, to my mother, a middle school mathematics teacher, and to me is "Telling is not teaching." In this volume, Susan Pirie (chap. 15) asks, "Is There Anybody Listening?" My family truism and Pirie's article bring to mind two questions posed twenty-five years ago by the author and educator John Holt (1970) in his book How Children Learn: *"Who needs the most practice talking in schools? Who gets it?" Would you share your thoughts on the value of listening and talking in mathematics classrooms?*

Pirie's question on *listening*, Holt's questions about *talking*, and your family's "truism" dramatically make the same points that I have espoused through the years. When teaching, we need to listen. We need to listen to what our students understand, to what they know, and to what they think about mathematics and mathematics learning. We need to listen for those silent questions that all our students have and establish an environment in our classrooms that encourages those silent questions to become expressed.

My own thinking on this topic has grown over the years. In an article I wrote for the *Arithmetic Teacher* (May 1988), I stopped with the thought about listening to the silent questioners. Now, I approach my students who are silent in a slightly different way. Using the *Professional Teaching Standards* as my guide, I now try to create an environment in my classroom where students know it is safe to make conjectures, to ask why, to explain their reasoning, to argue, and to resolve. Thus, I hope that not as many of my students will be silent questioners.

If we all look again at the *Professional Teaching Standards,* particularly the discourse standard—the teacher's role in discourse, the student's role in discourse, and the tools for enhancing discourse—we will see pictured a class quite different from the one in which teachers do all the talking and the students supposedly do all the listening. I am pleased to see that there are many articles in this yearbook that address the discourse standards. This is an area in which we all need to grow.

Many teachers are learning to be facilitators of classroom discourse, but there are still some who believe that successfully managed mathematics classrooms are those where the teacher talks and students speak only in response to questions to which the teacher already knows the answers. In the past, when we looked for indicators of successfully run classrooms, we looked for orderly chairs, straight rows, and teachers in charge. Are these still good indicators of successfully managed classrooms, or are there other indicators we should look for as we visit classrooms today? What I am asking is, How will the shifts in classroom discourse manifest themselves?

This is a good question with no simple answer because there will not be one description. Mathematics classes have often been described as being the same across the United States and Canada: check the homework, show how to do the next set of exercises, and have students begin the exercises. Unfortunately, this is an apt description of too many classrooms, although there are many notable exceptions. It is these notable exceptions that describe how the shifts in classroom discourse will be manifested.

In the introduction to the *Professional Teaching Standards,* five expectations were listed that answer your question well (NCTM 1991, pp. 3–4). They are simple, but powerful images:

- Students working together to make sense of mathematics
- Students relying more on themselves to determine whether something is mathematically correct
- Students learning to reason mathematically

- Students learning to conjecture, invent, and solve problems
- Students learning to connect mathematics, its ideas, and its applications

The question is, therefore, How can we help these expectations become part of our classes? The essence of all the standards is that they suggest, they probe, they encourage, and they require each of us with the help of our colleagues to develop our own interpretations. This is especially true of the Teaching Standards; our reflections on their guidance will help us bring about meaningful discourse in the classroom.

In this volume, Edward Silver and Margaret Smith (chap. 3) describe establishing "discourse communities" as making "a worthwhile but challenging journey." What do you see as some of the challenges schools face in moving toward "discourse communities"? And what, if the journey is successful, are some of the rewards that make it a worthwhile trip?

The journey has begun and we know where we are going, but we are still struggling with how to get there. We want to reach the point where each student has the opportunity to develop the mathematical understandings, skills, and dispositions needed for today and for the future. We want all teachers of mathematics to have the opportunity to develop professionally in a manner that will enable them to take their students on this journey and to have the support for this journey.

The journey will never be completed. We will not take all the same path. Some will travel along superhighways that are well kept and up to date. Others of us will face dirt roads or city streets. Some will follow lonely paths, and others will have crowded streets. Yet, no matter which path we take, we will each face barriers and often will have to take detours. Some days we will make little progress; other days we will sail along. Yes, some days we will be lost, looking for a new road sign or new direction. This is all part of change.

We did not become teachers because we thought the journey would be without adventure. Our rewards do not come at the end of the journey; they often come at the most unexpected times. We turn a corner, and there in front of us is that student whose eyes light up to say, "I understand." We reach an intersection, and our students make a report of their mathematical findings that clearly points to the next turn. We reach the end of a leg of our journey, and we realize how many more students are doing well in mathematics. A student returns from his first year of work to let us know how much he uses the mathematics he learned, or a student tells us she was accepted into an engineering program. These are our rewards.

Now, I have been talking about a broader journey, of which established discourse communities are just one part. These discourse communities give us company as we travel, but they also bring all the challenges that come with company. How do we communicate? How do we make everyone feel a part? How do we take care of everyone's needs? What happens when part of the group does not want to participate? How do we manage a crowd that is too large? This company brings all the rewards of sharing successes and failures to help us take the larger journey.

I worry, as I am sure you do, about the "superhighway trips" for some and the "dirt-road trips" for others. I know you were speaking metaphorically, but for me, your metaphor conjured up the deepening schism between the "haves" and the "have nots" about which Jonathan Kozol (1991) writes in his book Savage Inequalities. *I sincerely hope that one of the rewards of this journey we are taking will be opportunities for stakeholders to ask and answer questions like these: "Are all children being given the same opportunity to learn mathematics? If not, what can be done to ensure that they are given opportunities to demonstrate what they know and can do and not what they don't know?" And equally important: "How can we keep our children from being penalized by a system of 'high stakes' testing that has not found a way to remove its 'savage inequalities'?"*

Portia, I am concerned—as you are—with a system that wants to sort and seriate but not educate. It has always seemed ironic that a country filled with so many mathematical phobias could yet rely so heavily on a number for a grade or a simple numerical rating scheme for the answers to, or descriptions of, complicated phenomena. What a wonderful job we have done in mathematics in selling the power of a number! (I'm speaking "tongue in cheek" here.) What a difficult time future generations will have understanding how we ever accepted one number after we help our students understand mathematics, its uses, and its limitations!

I think it is time to communicate more of what we expect and to take seriously the standard from NCTM's *Assessment Standards for School Mathematics* that states that assessment should enhance learning. When we want to know what students are learning and not what they do *not* know, then we will turn to gathering our information from many sources and will factor in the "opportunity to learn" variables before drawing any inferences on entire populations of students.

The bolted-down seats that characterized classrooms in the first half of the century are no more. But as we move toward "discourse communities" and attempt to reform assessment practices, there is no reason to expect that the removal of society's "bolted-down thinking" has taken place. Mary, you and I both have heard from the critics and resisters of mathematics reform that "I didn't do all this 'communicating' when I was in school and I learned mathematics. I hated it, but I learned it." Given that this perspective is out there, what should we be communicating to stakeholders?

Portia, I love your "bolted-down thinking" metaphor. What images it brings to mind! I hear as often that "I learned it, but I don't use it." Perhaps that is the message we should be communicating—"Your child is learning the mathematics he or she will use." If we think that mathematics is a *language*, how do we learn a language? We talk, we listen, we read, we write. We build the concepts underlying the ideas so we can communicate with meaning. We build the skills that allow us to communicate with many others. If we think that mathematics is a *science,* how do we learn a science? We explore, we observe, we make conjectures, and we test those conjectures. We build those explorations and conjectures into applications and theories, and we use those applications and theories. If we think that mathematics is an *art,* then we will want to produce works to share with others. No matter what our view of mathematics, it requires communication. Where better to learn that communication than in our classes?

Mary, the most vocal of the stakeholders are parents. What do we say to concerned parents who are wondering if their children are really learning mathematics?

I think that we realize that every parent has a right to know both what mathematics we expect their children to know and be able to do and how their children are progressing toward these goals. I believe that first we need to use those communications skills we are developing in our changing roles of teaching and begin by listening to parents. Before answering the question of "What do we say?" we need to know what parents are thinking and their view of mathematics. Then we can tailor our answer to the parents and the child in question. To do so, our goals must be clear, and we must be confident—but not cocky—about our expectations.

We should have a clear message and be able to deliver it without an excess of jargon. Play on those aspects that we have in common with the parents. All of us want *all* students to do well, and parents, naturally, are most interested in their own children doing well. This fact needs to be recognized by both parties.

When I was in my preservice education courses, I remember reading the following anonymous quotation: "I saw tomorrow look at me through little children's eyes / And thought how carefully we would teach, if only we were wise." Frances Curcio and her coauthors (chap. 25) give us some insights into their preservice classrooms where teachers are learning to translate discourse practices into care-filled learning experiences for children. As a preservice educator returning to your classroom at Columbus College, what recommendations do you have for how preservice education may need to change as we understand more about the communication principles in the Standards *documents?*

First, I think those of us in preservice education need to embrace the views espoused in the *Professional Teaching Standards* as well as in the *Call for Change* of the Mathematical Association of America (MAA 1991). Probably the most important idea is that preservice teachers should experience mathematics being taught in the way they should teach.

Second, I think we must take seriously the continued professional development of teachers and rethink the continuum. We should do away with the idea of preservice/in-service teachers as a two-step function and make it a continuous function.

The need to look at professional development from the initial preparation through lifelong learning ranks with the need to take our message to parents. We are at a critical juncture with each.

I love the idea of thinking of preservice and in-service education along a professional development continuum. I am reminded of another continuum where dialogue is sorely needed: the formal education continuum. How might we establish better lines of communication across all school levels—preschool; elementary, middle, and secondary school; and collegiate?

This is certainly not a new problem. One new dimension is the trend toward more local and school automony and away from heavy central administration. We must look at ways to communicate without the advantage of having, in many systems, mathematics consultants or systems in place that encourage communication.

We need to approach from many directions the task of establishing better lines of communication. Professional organizations can assume an important role. For example, the Conference Board of Mathematical Sciences has appointed an education committee to bring educational issues to all the mathematical professional societies; the leaders of the American Mathematics Association of Two-Year Colleges, Mathematical Association of America, and NCTM have come together to address mutual concerns and activities; the Research Advisory Committee of the NCTM has begun a project of bringing teachers, mathematics educators, and mathematicians together around research issues; and other mathematical professional organizations have educational issues spanning the different levels on their agendas. The State Coalitions for Mathematics, Sciences, and Technology Education spawned by the Mathematical Sciences Education Board have brought together many of the diverse players in many of the states. We are beginning to see more joint conferences among the different professional groups, especially at the state level, and to recognize the need to talk and work with one another.

In addition to the work of the professional organizations, we each can do our part to find out what our colleagues are doing and let them know what we have been doing to improve mathematics learning in our classrooms. I have been fortunate to work with teachers at all levels. When I go from a second-grade classroom to one in algebra 2 and then to one in college geometry, I learn from each. Sometimes I learn about common problems and concerns all of us are facing, and sometimes I learn new ways to develop ideas, to manage classrooms, or to encourage discourse. At times, it is even mathematics that I learn. I also see things that disturb me, such as secondary school teachers doing the same mathematics that I have just seen being done in fifth or eighth grade—and being done at not much more of a sophisticated level. I see college classes for teachers that make no connections with what these students will be doing with their students. I see elementary school teachers who are not taking advantage of the possible mathematics found in activities their students are doing. Observing classes at levels they do not teach is only one way individuals can open the doors of communication. There are other ways. How often do we include middle school teachers on committees to develop high school curriculum guides? How often do we have sessions at conferences where elementary school teachers present to college faculty? How often in our secondary school methods courses do we include readings from the elementary school journals?

Curriculum guides and textbooks often do not clearly indicate the growth expected. For example, area formulas often are repeated at about the same level of complexity at sixth, seventh, and eighth grade, and again in geometry. The *Standards* documents envision growth, but we need to learn to communicate it more clearly in all that we do.

Ideally, better lines of communication could be established by individual teachers who assess what their students know and are able to do and plan accordingly; by teachers within schools working together on content and instructional issues across grades; and by school systems and other collaborations providing opportunities for teachers to work across school-unit levels on curriculum, to observe other levels, and to share ideas. Professional organizations, clearly, can furnish opportunities for communication and materials based on those discussions and can help us develop the respect for all parties involved.

The last section of this yearbook is devoted to the role of language in discourse. Zalman Usiskin (chap. 28) presents a rather compelling argument that "mathematics is a language" and as such should be experienced as natural language/literacy experiences. Rafael Olivares (chap. 27) gives us a framework for analyzing communication in mathematics for students with Limited English Proficiency (LEP). From some of the previous questions we can infer something of your view on "mathematics as a language," but what specific messages about language development in mathematics should we be sending to our students and teachers? Would you modify this message if I ask specifically about LEP students?

I agree with the point that mathematics should be experienced more as a natural language. I am struck with how we treat language in geometry throughout our schooling. This was brought home to me last spring when some preservice elementary–education students were presenting geometric activities to fellow students. Most began by saying, "We must first review the vocabulary." The activities involved very few specialized words; many activities were designed to build the vocabulary as students were exploring the geometric ideas. Consequently, geometry to many of these prospective elementary school teachers was predominately vocabulary.

As we move to more communication in our classrooms, I think we need to continue to examine the tools of communication that are special to mathematics, including our sophisticated symbol system. Students should become conversant with the system in a natural way, understanding the beauty and power of such a system. For example, they need to come to understand that our fraction notation has many meanings and is used in different ways to represent different relationships. This does not necessarily translate directly to a message for teachers and students, but I believe that we need more emphasis placed on research that looks at how students develop symbol systems. For example, how much practice is needed in manipulating symbols in a world in which technology will do it for you? How do we develop the understanding of what is actually happening?

Mathematics makes use of models, pictures, diagrams, and graphs to communicate. These are tools to make mathematics meaningful and to describe the mathematical ideas present. It is often easier for students to talk about physical or pictorial models than the abstract idea. Today, none of these tools needs to be static. Through technology we can bring the action into our classes. Students can manipulate objects in ways that were not possible earlier. For example, they can ask the computer for a view from above, from below, from the right. The message here is that we should remember that these tools are powerful means of communication.

Now, how would I have answered if you had asked about LEP students? I would probably have returned to some of the literature or talked to those who know much more about LEP students than I do. I need to learn more about LEP students and am looking forward to reading the article by Olivares. I am sure, however, that my message would echo those of many in this field who remind us that Limited English Proficiency does not mean Limited Mathematics Proficiency.

Technology is great, isn't it? We are able to have this conversation using multimedia technology (e.g., fax, e-mail). Ann Barron and Michael Hynes (chap. 17) suggest that "the future is now" when it comes to using multimedia technology to enhance communication in mathematics classrooms. Do you have any final thoughts about your vision for communication in mathematics classrooms twenty-five years from now? Do you realize that will be the year 2020? I hope our vision is as keen as the numerals representing that year suggest.

When I read your question, I could not help thinking about my father, who often told us about his school experiences. He vividly remembered the first day of the year 1919. He decided that very day that he would live to see the next time this kind of number pattern occurred, namely, 2020. His teacher, surely, must have influenced his love for numbers; I remember his excitement when the car's odometer reached any number with a pattern (and my embarrassment at thirteen when it reached 14141 for the second time in my life). His experience in helping other students in his one-room school must have influenced his decision to teach, to pursue his doctorate with William A. Brownell, and to devote his career to education.

I wonder what his vision of communication in the year 2020 would be. From his other stories, I believe that he would see the school as a place where all students were developing a love for learning, were being challenged by the unknowns and adventures that lay ahead, and were learning so they could become productive citizens. He would see caring teachers who were working together and sharing ideas, learning together new ways to help students learn, and enjoying their work. He would see teachers, administrators, and students working together to make the school a place of which they are proud. He would see communities supporting the schools and respecting the teachers and administrators. He would see continuing change and a striving toward making schools better for all students. His communication style would have been modeled on his belief that "action speaks louder than words."

Each of our visions probably differs from that of my father's. The day of the isolated teacher whose primary source of communication is the printed page is over. In 2020, there may not even be schools as places. Even today, technology has opened new ways of communicating. Students work with students in other classes across town or across the world. Distance learning connects students with other teachers and resources. Schools and colleges are linked so that prospective teachers can watch and interact with teachers and students. Teachers are sharing questions, materials, and successes through electronic mail. The transmission of graphics has developed to the point where we can share a variety of materials. Publishers provide materials that are truly interactive, both for the teacher and for the students. The speed and ways of communicating continue to change at an increasing pace.

No matter what 2020 brings (we can only hope that our vision will be recorded this perfectly), I hope that the essence of my interpretation of my father's vision will still be in place, for it parallels the vision of the *Standards:* a positive and productive education for all.

And now, no more words; it is time to act.

REFERENCES

Holt, John. *How Children Learn.* Harmondsworth, England: Pelican, 1970.

Kozol, Jonathan. *Savage Inequalities.* New York: Crown, 1991.

Mathematical Association of America. *A Call for Change: Recommendations for the Preparation of Teachers of Mathematics.* Edited by James R. C. Leitzel. Washington, D.C.: Mathematical Association of America, 1991.

National Council of Teachers of Mathematics. *Assessment Standards for School Mathematics.* Reston, Va.: National Council of Teachers of Mathematics, 1995.

————. *Professional Standards for Teaching Mathematics.* Reston, Va.: National Council of Teachers of Mathematics, 1991.

2

Diverse Communications

David Pimm

DIVERSE is a commonly heard word these days: diverse student popula-tions, diverse learning methods. It is rather less often that we hear about diverse teaching methods and the need to value them. We have even reached the stage where various descriptions of "the perfect lesson" have been offered, with the apparent presumption that it is to be continuously repeated day after day in all its balanced perfection. In this article, I intend to explore some possibilities for diverse mathematical activity and com-munication within a single classroom, one where mathematics is in focus. In addition, I argue for the importance of maintaining a rich diversity of styles and sources of classroom discourse within any individual teacher's repertoire.

Mathematics education seems particularly prone to the belief in the sin-gle new idea: do this (whether using calculators, teaching mathematics through problem solving, working collaboratively, stressing the basics, employing manipulatives, and so on), and *all* your mathematics teaching problems will be solved.

Such monomaniacal enthusiasms have marked the last ten, twenty, forty years of writing about the teaching and learning of mathematics. The ed-ucational system can seem so monolithic, so inert, that perhaps individual proselytizers need to be monomaniacs in order to shift the system's center of gravity at all. Yet there is also the danger of promoting a belief in the possibility of a quick fix, of instant renovation rather than thoughtful restoration.

Think for a moment about the diversity of ways of approaching the idea of number. Teachers can and do choose to offer students Cuisenaire rods, for instance, so as to supply a more tangible referent for number. Teachers can also create abacus-based activities so as to suggest a more transparent means of calculating. Teachers may work with students on di-rect finger calculations; it is easy to forget how much, for young children, the world is primarily a world of touch. There is a useful adage: "My fin-gers are an extension of my brain."

Teachers can and do choose to offer number-word games and rhymes — with no appeal to physical materials — where the task is almost entirely linguistic. They regularly present to students opaque Indian-Arabic numerals and the algorithms that this numeration system supports. Teachers can and increasingly do furnish students with electronic calculators.

Despite regular claims for something being *the* way to teach number, classroom decisions are seldom a choice of one method over another. In mathematics, we can offer Cuisenaire rods *and* finger complements *and* spoken numeral games. All these resources contribute to developing aspects of numbers, and all can be made to "hold" number and to carry out calculations in different ways. All contribute to the development of mathematical meaning. All, therefore, deserve time and a place in mathematics classrooms.

POSSIBLE WAYS OF WORKING MATHEMATICALLY

My particular area of interest is in language and mathematics; more specifically here, in situations and materials in the mathematics classroom and in the communication they can help generate. Mathematics is, in large part, brought into being by the *conversation* and is not available directly from the materials alone. Mathematics often starts with actions on objects but cannot stay there. Students must move from the direct exploration of the actual to the virtual exploration of the possible, from the heat of direct action to the gentle detachment of intense reflection—and back again. Active *and* passive, action *and* passion.

Nevertheless, actions serve as an initial setting and a focus for observation and discussion—a reason to observe and pay attention (because something is happening) and a reason to describe and comment (because it seems noteworthy and hence worth discussing with someone else). However, with physical objects and actions on them, there is a tension between reflection and immediacy. Instructions and other pedagogic devices are often required for interrupting the physical immediacy of the situation, so that the teacher may encourage reflection, prediction, and thought.

"Checking" in your head before physically checking the result on the apparatus is one possibility. With a paper-folding and prediction task, for instance, cutting and unfolding as quickly as possible is the simplest way for the student to find out what the shape will be *in this particular case*. But if the teacher's intention is to encourage geometric thought and the exploration of possibilities of what *could* be, then the teacher must try to slow the activity down and intervene in the student's thought processes to ensure time for reflection.

In this situation, as in many others, the teacher and the students are partial antagonists. The students (consciously or not) want to bring the task to a close as quickly as possible. Teachers want to prolong the imaginative exploration of the possible at the expense of the actual as long as it is feasible, and often without indicating that that is what they are doing. This will afford them the most opportunities for teaching.

The Russian psychologist El'konin puts it this way (cited in Davydov and Markova [1983, pp. 60–61]):

> An educational task differs fundamentally from other types of problems in that its goal and its result consist of a change in the acting subject himself, not in a change in the objects on which the subject acts.

Working with Manipulatives

School mathematical tasks can involve moving one's own body, manipulating objects such as cubes or rods, cutting paper with scissors, or folding cards. (There is an important distinction to be made between the mathematical task offered and the pupil activity generated in response to it. See Love and Mason [1992].) Tasks may involve more reflective settings such as contemplating and discussing patterns seen in a table (as in the hundred square, the multiplication table, a 5 × 5 magic square, or Pascal's triangle). But, however absorbing the student's activity may be, mere activity is not the end in itself.

Mathematics involves focusing on relationships between parts and wholes, exploring change and constancy, stressing this and ignoring that. Mathematical activity is the means to an end, to encountering some idea or isolating some property, to seeing or realizing that a certain something *must* happen or *cannot* happen.

However, if the activity generated by the task itself engages *all* the attention of the students working on it, the teacher's purpose in setting up the situation as a potentially *mathematical* experience may become diluted or even lost. What for the teacher is merely one of a collection of particular situations may for the students be the entire focus of their attention. With manipulatives, for instance, there is a danger that students may end up only manipulating the equipment.

A set of materials—physical apparatus—does not offer unmediated mathematical experience: by itself, it can neither contain nor generate mathematics. Only people can do this, with their minds, and it is a central part of a teacher's role as teacher to help students to become more able to do this for themselves. The poet T. S. Eliot writes of the possibility of having the experience but missing the meaning—a situation that may well abound in mathematics classrooms. In this sense, perceiving mathematics is fragile. One thing teachers need to do continually is attend to the mathematics in the situation and communicate where their own attention is—and not be too concerned at directing the student's attention there as well.

The most immediate and important manipulative is a student's own body. Caleb Gattegno's (1974) intensive and inventive work involving young children forming number complements on their hands draws extensively on their dynamic control over folding fingers up and down. To form "four," fold up *any* four fingers. To find the complement of four in ten, interchange folded down fingers with folded up ones. Rename each finger as ten ("-ty"). To find the complement of four-ty in ten-ty (a hundred), act likewise.

Whatever else, forming complements is something that children can *do.* They can work with their fingers and at the same time encounter multiplicity and equivalence (the myriad ways of showing four or seven, for instance); invariance (the complement in ten of any way of showing seven is always a way of showing three); the reciprocal nature of stressing and ignoring (turning fingers up into fingers down, and the converse, produces the complement); the condensation of number naming as well as its uniformity (choose a different unit for the fingers and produce complements in 100 or 1000). Finger work also offers an image for number that can be internalized directly through the digits. But all this will remain tacit, latent, unless it is brought to conscious, focused attention through teacher and class discussion.

It is not uncommon for younger students to be invited to exploit their own physical activity for mathematical ends. Janet Ainley (1988a) writes of a range of tasks involving games that make use of links between mathematics and movement in the elementary school, including working with muscle memory, rhythm, and counting. Although less widely used, similar resources do exist for mathematical tasks at junior high school or high school levels (e.g., Association of Teachers of Mathematics 1985; Bloomfield 1990).

One instance involved a class in which two children represented fixed points. The other members of the class were to place themselves twice as far from one of the "points" (children) as from the other. In doing this, some students commented that they could *feel* the constraints inside themselves. One teacher of eleven- and twelve-year-olds asked her class to describe images they had seen in a complex geometric poster (discussed further below). Toward the end of the lesson, she asked them to choose and fix their minds on one single image from among those they had seen. They then worked in groups of three or four to depict their selected visual image (for the rest of the class) physically and collectively using their bodies—and in silence. They communicated by a direct showing. The teacher reported being surprised at how comfortably and easily the groups worked together on this task.

However, such activity can be threatening for some adolescents. An experienced secondary school teacher, Anne Watson (1991, p. 28), in a review of Bloomfield's book, commented on her experiences:

> I was attracted to the title [*People Maths*] because I feel the need for people to be involved in mathematics, not just intellectually but with a range of senses and in situations which fit in with other aspects of their lives.... So why don't I use physical games as much as I could? My excuse is that youngsters of 13+ are often too embarrassed to move if they are not used to it. They are also heavily conditioned about what is, or is not mathematics. I have to work hard to get them to trust me and sometimes I do not get there. It depends what has gone on before.

With all these physical possibilities, the fact that the "object to think with"—the manipulative—is a part of the person doing the thinking can offer particularly direct access to the experience. However, the teacher may not readily notice if a student's attention is not focused on the mathematical possibilities, because so much physical action is smoothly—even subconsciously—integrated into a student's general mathematical functioning. Nor will such physical experience necessarily be transferred to other mathematically related situations that do not call for physical movement.

In many schools, three-dimensional wooden or plastic geometric shapes are available. One means of directing students' attention to certain properties of geometric forms is to focus on the tangibility of these shaped objects through the use of a "feely box" (Giles 1985). Such a box can be made from cardboard and has an open back and two holes for the arms of the person using it. All the members of the class can see in the back, with the exception of the person whose arms are through the holes.

That person however, can *feel* any shapes that are placed there. Students try to describe the solid shapes they can feel with their hands and perhaps identify them by name: having a reference set of possibilities on display may aid identification, depending on the level of challenge desired. (A similar task involves using a bag with wooden shapes in it, where no one can see which shape is being handled and described.) What do students focus on in their descriptions? What are the aspects of shapes that are most tangible? What descriptive language do they use? Can they identify an object from its feel alone? (For more on this, see Mason [1990].)

All the diverse examples so far have been particular. Are there some more general design principles for such tasks? The challenge above offers elements that can be widely used in designing classroom tasks. The first is focusing artificially on one of the human senses (with the feely box, that of touch) by excluding another (there, sight) and thereby producing a heightened sensitivity from throwing out the normal balance of the senses. Second, having to use words alone (for the names of shapes or properties) may focus students' attention on the need for, and usefulness of, having words for certain characteristics of a shape. It may also allow a teacher to seed the discussion with those words—without the words themselves necessarily being the main focus of the class.

Variants might include the following: sitting back-to-back or using a telephone to describe something to another person, who then must re-create it; describing something while sitting on your hands; saying what you have seen in a picture without pointing or touching. A useful separation (here of language and action, and controlling the latter by the former) can occur when one student acts as the "head" and a second as "the hands"; the head can only talk and give instructions, whereas the hands can only cut, draw, or paste according to what that student has understood. Again, it is important not just to offer the task—a period of discussion and comparison must follow, after which the students may have the opportunity to try again. (So often we hear "They can't do that" rather than "How can I help them get better at that?")

The constraints of these tasks are "artificial." But provided that the constraints are accepted by the children as "the rules of the game," then the constrained situation can offer a powerfully focused context for learning. If there is feedback and comparison, there is the opportunity for them to practice using language to point or gesture. Teaching is, by its very nature, artificial—an intervention in the learning process. One interesting classroom paradox, however, is that artificial teaching may nevertheless result in natural learning.

Working with Images

In addition to drawing on the actions performed on physical objects, mathematical thinking involves the repeated use of images: kinesthetic images (from bodily actions, such as the shrugging of shoulders), screen images (on calculators or computers), drawn static diagrams (in textbooks), and mental images themselves (in the student's mind's eye). How are we to communicate images and communicate about images?

Beeney et al. (1982) provide classroom accounts of the mental geometry activity that resulted from tasks involving turning words into images and back into words. Although words can serve to conjure images, because mental images are inherently private and personal, there is no direct way of offering them to others.

An effective way of creating images can be through the use of words alone—for instance, an imagining task such as "Think of a Picture." A teacher might begin by engaging students like this:

> Close your eyes. Imagine a square. Stretch it, shrink it, rotate it, move it around in your mind to get a sense of all the squares it might be. Bring it back now to the center of your "mental screen" and arrange it to be "square" to the picture, with a top, bottom, and two vertical sides.
>
> Add in a circle now, one whose diameter is smaller than the side of the square, and move it so it rests on the top of your square. Rock it back and forth until it starts to roll. Roll it all the way around the square, so that it is always touching the square, until eventually it comes back to where you started.
>
> Now select a point in your circle and imagine it leaving a trace as the circle rolls. (According to the desired level of mathematical difficulty, this point could be the center of the circle, somewhere on the circumference, somewhere in the interior of the circle, or even somewhere outside it—provided it always moves with the circle when the latter rolls.) Roll the circle all the way around the square again, focusing on your chosen point. Now, turn to the person next to you, sit on your hands, and describe only with words what your image was and the shape left by the trace of the point in your circle.

When shown a picture or a poster with mathematical potential and asked to say what has been seen, students often use a combination of words and gestures (perhaps pointing or touching or using direction of gaze) to direct the questioner's attention. Within the context of a mathematics lesson, the teacher may use different criteria to accentuate the mathematical potential of the task. Learning to speak like a mathematician involves becoming able to use language both to conjure and control personal mathematical images as well as to convey them to others.

One method for working on these themes follows. The class sits in front of a photograph or picture. Students then take turns sitting in a "hot seat" at the front of the class "saying what they see" to the rest of the class. The task requires "no pointing and no touching" so as to force the students' attention onto the adequacy of the description being given. The focus on the poster (fig. 2.1) or on the speaker allows the teacher to escape from the limelight and encourages the communication to flow from a student to his or her classmates rather than from the student to the teacher. The teacher can ask questions, and the students must work on refining their command of mathematical language to convey their desired meaning.

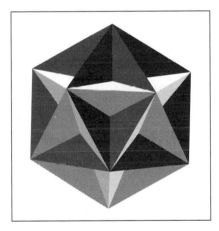

Fig. 2.1

There are a number of interesting aspects to this task. It is deliberately artificial. In everyday life, if you want to explain to someone something you have seen, it is normal and efficient to gesture as you describe. The constraints imposed in the task are part of what gives the lesson potential as a forum for mathematical learning. The focus is on the use of spoken language itself to stand on its own aside from accompanying physical gestures. Provided the students take on the "game" aspect (in the sense of a rule-governed task with a particular goal), then they are able to practice certain communication skills. These include, for the speakers, developing precision in describing what they see and, for the listeners, developing their ability to evoke images from words. An account of such a lesson with a junior high school class and a complex, multicolored poster as the geometrical focus, is given in Jaworski (1985). See also Pimm (1987).

One general mathematical purpose of the task above is to invite students to say what they see. The context of this task can help develop the student's ability to "point" with words alone, so as to focus the others' attention. This task also connects to the power of naming. If someone claims to be able to see a cube in a picture when others can't, then what the former speaker knows about that object can be brought into play in her descriptions, until others can perhaps come to see what was previously invisible to them. Students can thus realize that what is going on in their heads is hidden from others and that language is a powerful means of communicating such thoughts and images. Because the teacher also must visualize from the words being used by the students, there is more scope for "genuine" teacher questions (Ainley 1988b). The teacher desires to find out the "image that is in the student's mind" because she does not know and, in some fundamental sense, *cannot* find out in any other way.

CONCLUDING COMMENTS

In mathematics, there is always movement back and forth between the potential (the possible) and the actual. The question of alternative possibilities can be partially explored by looking at particular cases, but one important mathematical challenge lies in identifying *all* possibilities, eliminating impossibilities, and convincing others that all cases have been considered. Caleb Gattegno (1963) has made much of the distinction between *actual* (or *real*) and *virtual* actions. In particular, he claims that virtual actions extend the range of corresponding real actions yet still reflect certain constraints inherent in the original. He offers the following example:

> For instance, stringing beads is an action, while to imagine oneself doing this is, at first, to evoke the movement without actually carrying it out; to become aware of it as a possible action that can recur indefinitely is the virtual action which will serve as basis for the indefinite extension of addition of units. (P. 52)

More particularly, he proposes an interesting characterization of what it is to act as a mathematician:

> All those, then, who are capable of replacing actual actions with actions that are virtual and of contemplating the structures contained therein, act, when they do these things, as mathematicians. (P. 53)

This is one sense in which mathematics is necessarily of the mind; Gattegno identifies awareness of the structure of virtual actions as characterizing the mathematician.

The main elements of this article have been the need in mathematics both for action (actual, then virtual) and for reflection—and the importance of being able to communicate both. (For more on this, see Pimm

[1995].) In particular, very young children are capable of mental activity and imagery, and high school students can work with paper and scissors to achieve mathematical ends.

Communication about the mental realm where mathematics takes place is essential for others to gain access: both for information and for instruction. Such communication is also important as a central aid to students in formulating their own mathematical ideas. Students have a right to many diverse opportunities for encountering and engaging with mathematical ideas, and teachers bear the curricular obligation to furnish those opportunities as best they can.

REFERENCES

Ainley, Janet. "Maths in Motion." *Child Education* (September 1988a): 33–35.

———. "Perceptions of Teachers' Questioning Styles." In *Proceedings of the Twelfth Psychology of Mathematics Education Conference,* edited by Andrea Borbás, pp. 92–99. Veszprém, Hungary. OOK Printing House, 1988b.

Association of Teachers of Mathematics. *Away with Maths.* Derby, England: Association of Teachers of Mathematics, 1985.

Beeney, Roger, and others. *Geometric Images.* Derby, England: Association of Teachers of Mathematics, 1982.

Bloomfield, Alan. *People Maths.* Cheltenham, England: Stanley Thornes, 1990.

Davydov, Vasilii, and Aelita Markova. "A Concept of Educational Activity for Schoolchildren." *Soviet Psychology* 21(2) (1983): 50–76.

Gattegno, Caleb. *Modern Mathematics with Numbers in Colour.* Reading, England: Educational Explorers, 1963.

———. *The Common Sense of Teaching Mathematics.* New York: Educational Solutions, 1974.

Giles, Geoff. *Feely Box.* Diss, England: Tarquin Publications, 1985.

Jaworski, Barbara. "A Poster Lesson." *Mathematics Teaching* 113 (December 1985): 4–5.

Love, Eric, and John Mason. *Teaching Mathematics: Action and Awareness.* Milton Keynes, England: The Open University, 1992.

Mason, John. *Shape and Space.* Milton Keynes, England: The Open University, 1990.

Pimm, David. *Speaking Mathematically: Communication in Mathematics Classrooms.* London, England: Routledge, 1987.

———. *Symbols and Meanings in School Mathematics.* London, England: Routledge, 1995.

Watson, Anne. "Getting in Touch." *Mathematics Teaching* 135 (June 1991): 26–27.

3

Building Discourse Communities in Mathematics Classrooms:
A Worthwhile but Challenging Journey

Edward A. Silver

Margaret S. Smith

COMMUNICATION and discourse are central to the current vision of desirable mathematics teaching (National Council of Teachers of Mathematics [NCTM] 1989, 1991). This view of mathematics teaching is quite different from the way traditional mathematics instruction has been designed and delivered. In traditional mathematics instruction, the role of the teacher is essentially to transmit knowledge to, and validate answers for, students, who are expected to learn alone and in silence. In contrast, according to the reform vision of mathematics classrooms, the role of the teacher is diversified to include posing worthwhile and engaging mathematical tasks; managing the intellectual activity in the classroom, including the discourse; and helping students to understand mathematical ideas and to monitor their own understanding. Students are expected to engage in doing mathematics while participating actively in a "discourse community." Thus, the new role envisioned for mathematics teachers is one intimately tied to issues of communication.

Interest in communication is both more widespread and more central to mathematics education reform efforts than ever before. Nevertheless, recognizing the centrality of communication as an issue for mathematics education is necessary but not sufficient to ensure a higher frequency of

The preparation of this paper has been supported by a grant from the Ford Foundation for the QUASAR project. The opinions expressed herein are those of the authors and do not necessarily reflect the views of the foundation. The authors are grateful to the teachers in the QUASAR project, who have generously given permission for us to tell their stories and who have bravely encouraged us to write not only about their triumphs but also about their struggles in order to offer encouragement and assistance to other teachers.

communications-rich mathematics teaching than has been typical of mathematics teaching in the past. Even when there is a high level of interest and commitment to communication as a feature of mathematics instruction, many teachers may struggle with the challenges arising from implementing these beliefs in classrooms.

TRAVELING A PATH TOWARD INCREASED COMMUNICATION IN MATHEMATICS CLASSROOMS: MEETING SOME CHALLENGES ALONG THE WAY

It should not be surprising that the process of creating mathematical discourse communities is a challenge, since it is a complex, multifaceted undertaking. For example, the *Professional Teaching Standards* (NCTM 1991) identifies six standards for teaching mathematics, of which one is "the teacher's role in discourse." To elaborate this standard, many aspects of a teacher's role in discourse are identified (p. 35):

- [P]osing questions and tasks that elicit, engage, and challenge each student's thinking;
- listening carefully to students' ideas;
- asking students to clarify and justify their ideas orally and in writing;
- deciding what to pursue in depth from among the ideas that students bring up during a discussion;
- deciding when and how to attach mathematical notation and language to students' ideas;
- deciding when to provide information, when to clarify an issue, when to model, when to lead, and when to let a student struggle with a difficulty;
- monitoring students' participation in discussions and deciding when and how to encourage each student to participate

The *Professional Teaching Standards* presents a wonderful image of a "last stop" on a long journey—classrooms as mathematical discourse communities—but it says little about the various paths along which teachers might travel to arrive there or about the challenges they may encounter along the way. The challenges may lead teachers to take "side trips" or make "rest stops" along the path; these diversions should be expected and not viewed as failures. Just as it is unrealistic to expect a polished solution to a complex mathematics problem when one first thinks about it, it is also unwise to expect perfection from one's fledgling attempts to create communications-rich mathematics classrooms.

Our comments and suggestions regarding useful ways to think about the journey toward classroom discourse communities are based on observations drawn from the QUASAR project. (QUASAR [Quantitative Understanding: Amplifying Student Achievement and Reasoning] is a national educational reform project aimed at fostering and studying the development and implementation of enhanced mathematics instructional programs for students attending middle schools in economically disadvantaged communities. There are currently six QUASAR school sites dispersed across the United States, and they represent a good cross section of urban middle schools in the country.) For the past several years, the mathematics teachers in the project have been collaborating with at least one mathematics educator from a local university to enhance their schools' mathematics instructional program through an emphasis on mathematical thinking, reasoning, and problem solving. In keeping with the *Curriculum and Evaluation Standards* (NCTM 1989), the kinds of instructional practices that many teachers have been attempting to establish in QUASAR classrooms emphasize engaging students with challenging mathematical tasks, enhancing students' levels of discourse about mathematical ideas, and involving students in collaborative mathematical activity (Silver, Smith, and Nelson 1995; Stein, Grover, and Henningsen in press).

Since the beginning of the project in the fall of 1990, many teachers have traveled quite far down a path toward the establishment of mathematical discourse communities in their classrooms. Rather than focusing on the wonderful examples of communication–rich mathematics teaching that could be found in their classrooms today, however, we shall discuss instead some challenges these teachers faced during their travel along the way. In this way we hope that their struggles with these challenges will be useful to others who journey along a similar path. Our focus here is on issues associated only with oral communication; limited space prevents a discussion of the challenges associated with written communication.

The Challenge of Getting Students to Participate

Teachers have found that a critical aspect of building classroom learning communities in which students are willing to engage in investigation and discourse is the creation of an atmosphere of trust and mutual respect (Silver, Smith, and Nelson 1995). Unless the classroom environment is safe for thinking and speaking, students will be reluctant to propose their tentative ideas and hypotheses, to question assertions that are puzzling to them, or to share their alternative interpretations. Since students may arrive at middle school with the assumption that it is acceptable to criticize fellow students for doing something they would characterize as "dumb" or "stupid," teachers must establish norms for discourse and social interaction in inquiry-oriented communities. Students must be encouraged to question one another's ideas and assertions, yet teachers must demand that students respect one another as persons. In the classroom communities teachers seek to build, criticizing someone's ideas is acceptable but criticizing the person is not. For many teachers, this is seen as being critically important in helping students develop their self-esteem. With middle school students, who are generally self-conscious and socially anxious, the building of a classroom atmosphere of trust and respect is no small challenge. Nevertheless, the goal of building an atmosphere of trust and mutual respect can sometimes conflict with the goal of having rich mathematical discussions, especially early in the journey toward mathematical discourse communities.

An illustration of the challenge faced by teachers in this regard is taken from one classroom early in the project. In this sixth-grade classroom, many of the students spoke English as a second language, and many were self-conscious and somewhat nervous about talking in front of the class. The first unit of study in the curriculum dealt with data representation. After students had become familiar with several different techniques for representing data (e.g., bar graphs, line graphs, pictographs), their teacher, Ms. Arnold, asked them to collect their own data in a survey of classmates. The theme of the survey was, "What is your favorite _____?" and students were free to select whatever topic interested them (e.g., icecream flavors, television shows, musical groups, types of music). Students worked in groups of four to collect data from classmates and others in the school, after which they were asked to make a graphical display of their findings. Once the graphs were completed, Ms. A asked each student group to select a spokesperson, who would make a presentation to the class. After each presentation, students in the class were invited to ask questions of the spokesperson in order to clarify their understanding.

In addition to satisfying her mathematical goals, Ms. A intended to use this project as an experience that would lead to students' increased comfort with presentation and public discussion. It was clear that she made a good choice, since the activity was accessible to the students, they were highly engaged with the survey and data representation, and they displayed a great deal of pride and ownership in their graphical displays. Ms. A gave clear ground rules for the discussion, including making it clear that no disrespect or ridicule would be tolerated. In this regard, the ensuing discussion was a great success. Students listened carefully to each presentation, and in their questions and comments they were quite respectful of one another. However, their questions tended to deal with nonmathematical aspects of the process (e.g., "How did you decide which TV shows to include?" "How did you divide the work?" "How long did it take to design the graph?").

Ms. A had accomplished much in the way of creating an atmosphere in which students learned to respect one another's ideas and to participate in discussions related to mathematical tasks. For more than one full class period, the classroom was rich in discussion, moving freely back and forth between English and Spanish, as students expressed their ideas and explained their work. Nevertheless, Ms. A was challenged by the many demands of the setting. Although there were important mathematical issues that could have been addressed (e.g., why a certain graphical form was selected for the data being presented; how matters of scaling entered into the interpretation of the graphs created by some groups), these issues were virtually ignored in the students' questions and comments. Ms. A did attempt to direct attention to some of these issues, but unfortunately the students did not respond in ways that supported her attempts to focus on the embedded mathematical ideas. She did not press the students on these matters. By allowing the class discussion to proceed as it did, Ms. A built a sense of mutual trust and safety in which the students could participate in public discussion; in so doing, however, she allowed many mathematical issues to go unexplored.

While trying to establish a discourse community, a teacher may legitimately decide that pressing students for more discussion of mathematical ideas must wait until a later time. Even teachers who want their students to understand that mathematical ideas are the topics most valued in discussions in their classroom may decide it is prudent to move toward that goal one step at a time. If one sees the development of classroom discourse communities as a journey, then it seems reasonable to begin in a safe, possibly nonmathematical space, in which students may initially be more comfortable, and then move gradually to settings in which the mathematical ideas are salient in the discussion.

As students' confidence and comfort grows, the teacher must ensure that the mathematics does not get lost in the talk and that progress is made along the path not only toward a real mathematical discourse community but also toward the increased mathematical proficiency of all students. After students become sufficiently comfortable with themselves as participants in classroom discourse, it is essential that their continued growth in self-confidence result from an actual increase in being able to do mathematics.

The Challenge of Centering Discourse on Worthwhile Mathematical Ideas

In addition to the challenges inherent in encouraging students to participate in classroom discourse, teachers face additional challenges as they seek to center that discourse on worthwhile tasks that engage students in thinking and reasoning about important mathematical ideas. According to the *Professional Teaching Standards,* "good tasks are ones that do not separate mathematical thinking from mathematical concepts or skills, that capture students' curiosity, and that invite them to speculate and to pursue their hunches" (1991, p. 25). Worthwhile mathematical tasks often lend themselves to multiple solution methods, frequently involve multiple representations, and usually require students to justify, conjecture, and interpret. Thus, these tasks can furnish rich opportunities for mathematical discourse. Nevertheless, it is not simple for mathematics teachers to integrate such tasks into their classroom instruction, especially if they have been accustomed to using more traditional exercises to be solved by applying rehearsed procedures. It may take teachers some time to shift from having students work alone and in silence to having them work collaboratively on worthwhile tasks that lead to discussions of alternative approaches and justifications of solutions. There are challenges involved in using worthwhile mathematical tasks and centering the discourse on them, as can be seen in the following episode.

Students in Mr. Johnson's seventh-grade class were completing a task that required them to express various ratios, presented in a variety of formats (e.g., 4:12, 15/25, 1/5 to 1/2, and verbal problems), in simplest terms. Mr. J modeled the solution of a few examples, and then, in order to have students experience mathematics as a collaborative activity, he encouraged his students to work and talk with one another in small groups. As they worked, he circulated around the room, stopping periodically to ask questions. When most of the class had completed the assignment, Mr. J orchestrated a large-group discussion to review the solutions. To provoke thoughtfulness in the classroom discourse, Mr. J frequently asked students questions about their answers (e.g., "How do you know?" "Does it make sense?" "Can you justify your reasoning?").

There are some laudable aspects of Mr. J's teaching evident in this episode. For example, he asked good questions that were directed at eliciting student thinking, and he furnished opportunity for students to share their solutions and their reasoning publicly. He clearly had progressed in satisfying the demands of "listening carefully to students' ideas; asking students to clarify and justify their ideas orally" (NCTM 1991, p. 35). Nevertheless, the classroom discourse in this case focused on how each simplest-terms ratio was obtained by applying a procedure, such as "multiplying the means and the extremes" or "dividing the numerator and the denominator by the greatest common factor." Because each ratio was considered in isolation, generalizations were not made about important ideas associated with ratio and proportion. Although Mr. J clearly understood the importance and value of asking students to share their thinking and justify their answers, the discourse centered only on the thinking associated with "giving the answer" and "telling how you did it."

This episode suggests that teachers may be challenged by the demands of "posing questions and tasks that elicit, engage, and challenge each student's thinking" (NCTM 1991, p. 35). In this episode, Mr. J's choice of task and type of questioning limited what students could discuss in class. For Mr. J, and for many other teachers as well, it is natural for the journey toward mathematical discourse communities to begin with the kinds of tasks he had always used with students. Unfortunately, many tasks used in traditional mathematics instruction do not lend themselves to rich discourse. Through reflection on episodes such as this one and through consultation with colleagues, teachers can see the limitations of such tasks and become better able to select and use tasks that give greater opportunity for the exploration and discussion of important mathematical ideas. Although selecting and using a worthwhile mathematical task is desirable, the use of such a task does not alone ensure that the classroom discourse will be centered on important mathematical ideas. Even with a good task, teachers are faced with challenges as they seek to move students' explanations beyond procedural recitation and to probe for justifications that are based on making sense of a complex situation. An additional challenge is associated with departing from the habit in traditional mathematics instruction to simplify a problematic task by explicitly specifying a procedure to be used, thereby depriving students of opportunities to think through the problem on their own.

As we have seen in the situation described above, even if the tasks used in the classroom are related to interesting and important mathematical concepts and even if they are open to multiple routes of exploration, they can nevertheless be ineffective in promoting learning if students and teachers are not able to use them in an open way to explore and then discuss the emerging mathematical ideas. In order for tasks to have the desired impact for engaging and challenging students, teachers need to implement them in a manner that ensures that students understand the task and are able to make progress toward its solution. To be successful in implementing cognitively demanding tasks in the intended manner, teachers may need to use a combination of pedagogical strategies, including modeling high-level performance, supporting students by performing part of the task and leaving the remainder for them to do, giving students sufficient time to generate and explore their own ideas, encouraging student self-monitoring, and consistently communicating the need for explanation, meaning, and understanding (Stein, Grover, and Henningsen in press).

SUPPORTING TEACHERS TO MEET THE CHALLENGES

As we have seen, creating mathematical discourse communities is a challenging undertaking for teachers. Along the journey toward new forms of mathematics instruction, teachers are confronted with dilemmas and challenges. Ms. A clearly struggled with the challenge of accommodating and balancing apparently conflicting demands found in reform descriptions of classroom discourse communities—a conflict represented in the tension between helping students feel safe and comfortable participating in classroom discussion and orchestrating the discourse so that it focuses on mathematics and reasoning. And Mr. J was challenged by the interplay between the new forms of instructional practice he sought to establish and his experience with more traditional tasks and pedagogy.

The creation of discourse communities in mathematics classrooms is especially challenging at this time because most teachers lack personal experience with such environments. Most teacher education programs do not furnish prospective teachers with extensive experience with mathematical discourse, nor do most graduate-degree programs for teachers. In-service workshops are also an insufficient source of such experience. Moreover, since most teachers have learned the mathematics they know in traditional classrooms, they are being asked to create instructional environments with which they have had little direct experience either as teachers or as learners. Helping teachers to move away from a pedagogy of isolation and recitation and toward a form of instruction rich in collaboration and communication is likely to require new forms of experience and support.

One aspect of needed experience is that of learning mathematics in a manner that emphasizes discourse. This is important not only because it will provide an appropriate personal learning experience for teachers but also because it will provide an opportunity for teachers to see, hear, debate, and evaluate explanations and justifications of mathematics itself. This seems crucial, since the challenges of creating and managing mathematical discourse communities are intertwined with limitations in teachers' subject-matter knowledge and limitations associated with the ways in which they have learned the formal mathematics they know.

To be successful in teaching mathematics in the manner suggested by the *Professional Teaching Standards*, teachers need broad, deep, flexible knowledge of content and pedagogical alternatives. Moreover, they need to be capable of modeling reasonably good mathematical thinking and reasoning as they engage in "deciding what to pursue in depth" and "when to provide information, when to clarify an issue, when to model, when to lead." We have seen in the project that the acquisition of greater mathematical proficiency is an important contributor to teachers' success in moving toward the creation of authentic mathematical discourse communities in their classrooms. Although not sufficient by itself, a rich, personal experience with learning mathematics in this way, with the resultant increase in mathematical competence and confidence, can help teachers manage some of the challenges identified in this paper.

Another aspect of needed support parallels a need described above for students. Teachers, too, need to feel that they are operating in safe, supportive environments. As they struggle to change the atmosphere in their mathematics classrooms and the learning opportunities for their students, it is essential that they feel encouraged and supported by colleagues and supervisors in their efforts. The ability of teachers to deal with the many challenges and dilemmas suggested in this paper may depend to a great extent on the ability of colleagues to form supportive, collaborative communities of practice in which the discourse of mathematics teaching occurs.

Within the project, we have clearly seen the value of *community* in giving support and assistance to teachers, like the ones in this paper, as they meet challenges and make progress toward the establishment of mathematical discourse communities in their classrooms (Stein, Silver, and Smith forthcoming). The power of collegiality has also been recognized by others. For example, in describing one teacher's journey toward a form of mathematics instruction rich in communication and inquiry, Silver, Kilpatrick, and Schlesinger (1990) also referred to the role and value of collegial interaction. And Romagnano (1994) points to the value he derived from working with a colleague over time to "wrestle with" many dilemmas they encountered as they sought to change the way mathematics was taught and learned in their classrooms.

As we have seen, there is no reason to expect the journey toward mathematical discourse communities to be simple or brief. Since it is so important for the mathematics education of current and future generations of students, we can only hope that many teachers will not only begin but also persist in making the challenging but worthwhile journey toward the creation of mathematical discourse communities. In our view, if teachers are committed to the value of making the journey, if they understand that challenges and dilemmas await them along the way, and if they see themselves joined with colleagues locally and in the larger mathematics education community who are engaged in a similar journey, they can be successful in staying on course toward their goal.

REFERENCES

National Council of Teachers of Mathematics. *Curriculum and Evaluation Standards for School Mathematics.* Reston, Va.: National Council of Teachers of Mathematics, 1989.

————. *Professional Standards for the Teaching of Mathematics.* Reston, Va.: National Council of Teachers of Mathematics, 1991.

Romagnano, Lew. *Wrestling with Change: The Dilemmas of Teaching Real Mathematics.* Portsmouth, N.H.: Heinemann, 1994.

Silver, Edward A. "Moving beyond Learning Alone and in Silence: Observations from the QUASAR Project concerning Some Challenges and Possibilities of Communication in Mathematics Classrooms." In *Innovations in Learning: New Environments for Education,* edited by Leona Schauble and Robert Glaser. Hillsdale, N.J.: Lawrence Erlbaum Associates, forthcoming.

Silver, Edward A., Jeremy Kilpatrick, and Beth Schlesinger. *Thinking Through Mathematics: Fostering Inquiry and Communication in Mathematics Classrooms.* New York: College Entrance Examination Board, 1990.

Silver, Edward A., Margaret S. Smith, and Barbara S. Nelson. "The QUASAR Project: Equity Concerns Meet Mathematics Education Reform in the Middle School." In *New Directions in Equity in Mathematics Education,* edited by Elizabeth Fennema, Walter Secada, and Lisa Byrd Adajian. New York: Cambridge University Press, 1995.

Stein, Mary Kay, Barbara W. Grover, and Marjorie Henningsen. "Enhanced Instruction as a Means of Building Student Capacity for Mathematical Thinking and Reasoning." *American Educational Research Journal,* in press.

Stein, Mary Kay, Edward A. Silver, and Margaret S. Smith. "Mathematics Reform and Teacher Development from the Community of Practice Perspective: An Example from the QUASAR Project." In *Thinking Practices: A Symposium on Mathematics and Science Learning,* edited by James Greeno and Shelly Goldman. Hillsdale, N.J.: Lawrence Erlbaum Associates, forthcoming.

4

Meaningful Communication among Children: Data Collection

Susan Folkson

IN OUR technological society we are constantly being bombarded with information from every medium. One of the most efficient and concise ways for the media to communicate data is by using graphs. As adults, we read these graphs and make sense out of them. As educators, we teach our students to collect and read data using the same types of conventional graphs. The problem, as I learned by observing and listening to my kindergarten students, is that these conventional graphs are created by adults. They do not represent young children's thinking.

I used to do a lot of graphing activities in my kindergarten classroom. A vertical or horizontal picture graph would be designated—by me, of course—on a large piece of chart paper. Among the first things we would graph were our favorite colors. Other graphs would represent data about eye color and month of birth. After a graph was completed, I would pose my standard questions: "Which color is most liked?" "How many children like blue?" "Which color is more liked—red or pink?" "How do you know this?" When my students responded to these questions with the answers I was looking for, I assumed that they understood collecting and compiling data. It wasn't until I participated in a project focusing on how young children collect and organize data that I realized that I wasn't approaching data collection in the same way in which I taught other areas of mathematics.

For many years, I have been interested in students' thinking about number and problem-solving strategies for addition, subtraction, multiplication, and division. I have never imposed a way to solve these types of problems. I

This project was funded in part by a grant from the Professional Staff Congress of the City University of New York (PSC-CUNY) Faculty Research Program, awarded to Frances R. Curcio, Queens College. The results and opinions expressed in this work are the author's and do not necessarily represent those of the PSC-CUNY program. The author would like to acknowledge Karen Mertz for her assistance in collecting the data for this project.

let students develop strategies that are meaningful to them. In my classroom students are given adequate time to solve problems. I then invite them to share their ideas with one another. When students are allowed to share, they learn from one another and I learn more about their thinking, so I can appropriately develop my curriculum to meet their individual needs.

COLLECTING MEANINGFUL DATA

My challenge, then, was to approach data collection in the same developmental way. The first thing I had to figure out was what kind of data should be collected. After all, who really cares if more students like red than pink (Russell 1988)? I had to develop a purposeful topic that would require students to read the data they collected to find out useful information. Students in a kindergarten classroom constantly need their shoes tied! It would be very useful for the children to know who in the class can tie shoes. The teacher selected three students—Jacci, Jonathan, and Olivia—to collect the data. The three appointed students decided that they would write the name of each child in the class and then put a *yes* or a *no* next to the name. As you might predict, this task became tiring very quickly. Jonathan lasted only about five minutes. Jacci persevered for ten minutes. Olivia made sure she had every name in the class. She worked for thirty minutes to collect and record all the information (see fig. 4.1). When compiling the data, the students decided that it was unimportant to include the names of those who couldn't tie their shoes. Olivia made a list of the names of students who had a *yes* after their names (see fig. 4.2). They then decided where to hang the list in the classroom.

Fig. 4.1. Olivia's data for "We Can Tie Shoes"

The class continued the data-collection activities in small groups of students. The results from some topics yielded helpful information. After I

Fig. 4.2. Olivia's classroom display "We Can Tie Shoes"

found out "the most liked cookie," I was able to buy that type as a special treat for the class. The children were told that they could get a new board game, so we figured out which game to buy by collecting data about favorite board games. This kind of data did not give information that was useful to the students; it was of use to an adult.

ORGANIZING DATA MEANINGFULLY

I decided to ask the students what information they were interested in finding out. Since the practicality of the information was not a criterion, I allowed the students to select any topic that interested them. Christina, Jacci, and Olivia decided to collect data on the different languages spoken in the class. They knew from past experience that a lot of writing

would be involved, so they decided not to record the answer of a student who responded "only English." Christina and Jacci still found that the task required too much writing. The three students came up with the idea of making boxes. In each box, they wrote the name of a language that was mentioned during data collection. One box was labeled "A Little Bit of English and a Lot of Korean." Other boxes were labeled "Spanish," "Italian," "Indian/Persian," "Pakistani," and so on. In each box were listed the names of the students who speak the language. This activity showed how these students began to organize data in a way that was meaningful for them. They explained their thinking to the class (see fig. 4.3). Throughout the semester students continued to decide on data topics and presentation formats.

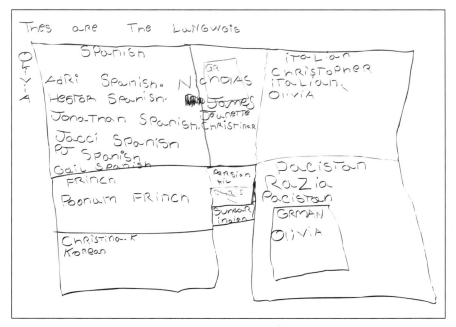

Fig. 4.3. Christina, Jacci, and Olivia's display "These Are the Languages"

In June, I asked the students if they were interested in knowing how many children were going to stay at our school and how many students were going to different schools. They were very curious. I asked the students how we could get this information. It turned out that each child verbally told the class the school he or she was going to attend in the fall. To figure out how many children would be going to each school, they decided to break up into groups. Gail, Jonathan, Nicholas, and Christina decided to record the data on paper. Using the idea previously developed by the students, Christina, Nicholas, and Gail decided to use boxes. They sorted the students by school and then labeled each group (see fig. 4.4). Jonathan's data were not as well organized. On a sheet of paper, he randomly wrote the name of every child and the school she or he would be attending. Jonathan then drew a line to connect each name to the school (see fig. 4.5). From this activity I concluded that Jonathan's organization skills were still developing.

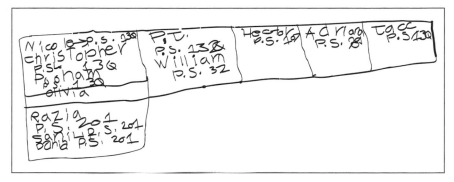

Fig. 4.4. Christina, Nicholas, and Gail's display of data on the schools their classmates will be attending

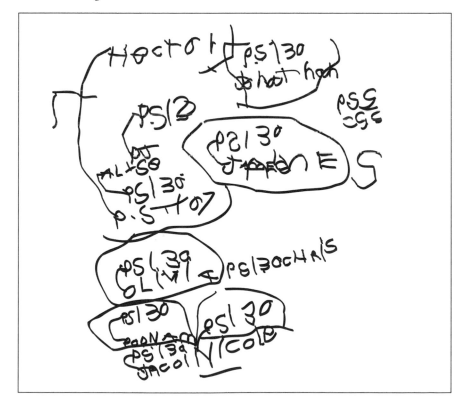

Fig. 4.5. Jonathan's organization of the data on the schools the students will be attending

CONCLUDING COMMENTS

The children were given many opportunities to collect and organize data. They worked in small groups and as a whole class. They listened to one another's ideas and adopted some of them as their own. As they progressed, the compilation of their data generally became more efficient and better organized. Not once, however, did a child choose to organize (or think to organize) his or her data in the form of either a vertical or a horizontal bar graph. After years of imposing those structures on students, I realized that I wasn't communicating very much to them after all. They learn a great deal more by constructing their own presentation formats. I am thankful that I've opened my eyes and ears to my students. They teach me so much!

REFERENCE

Russell, Susan Jo. "Who Found the Most Shells? (Who Cares?)" *Elementary Mathematician* (1988): 4, 9.

5

The Link Sheet:
A Communication Aid for Clarifying and Developing Mathematical Ideas and Processes

Mal Shield

Kevan Swinson

Mathematics classrooms are busy places where because of the many constraints operating, student–student and student–teacher verbal communication is often restricted. The benefits gained from the use of written communication in mathematics have received considerable attention in recent years. Writing about mathematics has been found to aid students' learning in many ways by encouraging reflection, clarifying ideas, and acting as a catalyst for group discussion (Rose 1989; Miller 1991). Writing in mathematics helps realize one of the major goals in teaching, namely, that students understand the material being studied.

When students' learning experiences are structured in such a way that connections and associations are made among different representations of new ideas, learning becomes more meaningful and useful because of the connected network of ideas so created. Such writing activities as explaining and describing assist students in making these connections and also allow teachers to "view" students' concept development (Borasi and Rose 1989). The idea of connecting procedures and concepts with different representations and prior knowledge has been described as *elaboration* (Swing and Peterson 1988). Borasi and Rose (1989) have noted that students who generate elaborations, such as visual images, as they read display a deeper understanding and a better memory of the material being read. Moreover, students who elaborate as they learn are better able to apply the ideas in problem situations.

WHAT IS A LINK SHEET?

With this notion of elaboration in mind, we sought a way to help students to elaborate on, and communicate about, the mathematical ideas and procedures they were learning. After a number of trials, we devised the "link sheet." It consists of a page divided into four rectangular sections, one each for a mathematical example; an everyday example; a diagram, picture, or graph; and the student's explanation, as shown in figure 5.1. This activity encourages the student to connect several representations of a mathematical idea and to communicate its meaning to others.

We have used the link sheet as a summarizing activity at the completion of a topic or as a diagnostic tool to indicate areas of weakness or the level of a student's conceptual development. Using different representations in the four boxes, the student gives examples of the specified word, symbol, or procedure. The examples in the boxes may or may not be interconnected. Students commonly illustrate the mathematical example in the "Diagram/picture/graph" box, whereas the everyday example often stands alone. In the box for "My explanation," the students generalize their thoughts. This activity has been used with classes at both the secondary and college levels. It may be completed with varying amounts of assistance from the teacher, depending on the experience of the students. The activity can also form the basis of a group discussion, either at the time the students are working on the sheet or after the students have completed the sheet independently.

A Summarizing Activity

One student's work in an eighth-grade class (ages 12–13 years) is shown in figure 5.1. This class had been working on a unit in algebra for several weeks when the activity took place. These students had used the link sheet only once before, so the responses were developed in a whole-class discussion. The students provided mathematical examples of simplifying algebraic expressions. This task presented few difficulties, since they had been working on these types of exercises in the previous lesson. The students suggested the example of the "cookie drive," since this fund-raising activity was taking place in the school at the time. The teacher was able to direct the discussion so that expressions could be used to show the amounts of money raised. Although not completely successful, the example gave some students an appreciation of an application of expressions. For the "Diagram/picture/graph" box, the students required prompting to recall the work they had done with materials two weeks earlier. The exercise was therefore valuable because it recalled an association that most students had apparently forgotten. The students completed the "My explanation" box themselves. This explanation also proved difficult for many of them and provided useful feedback to the teacher about what the students thought they were actually doing.

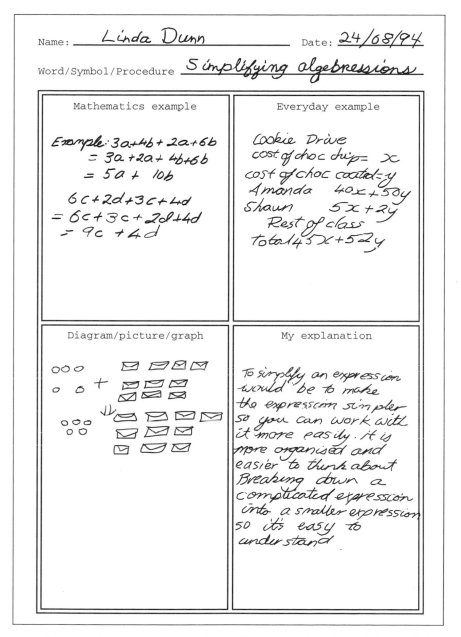

Name: _Linda Dunn_____   Date: _24/08/94_

Word/Symbol/Procedure _Simplifying algebressions_

Mathematics example	Everyday example
Example: $3a+4b+2a+6b$ $= 3a+2a+4b+6b$ $= 5a+10b$ $6c+2d+3c+4d$ $= 6c+3c+2d+4d$ $= 9c+4d$	Cookie Drive cost of choc chip = x cost of choc coated = y Amanda $40x+50y$ Shaun $5x+2y$ Rest of class Total $45x+52y$

Diagram/picture/graph	My explanation
	To simplify an expression would be to make the expression simpler so you can work with it more easily. It is more organised and easier to think about Breaking down a complicated expression into a smaller expression so it's easy to understand.

Fig. 5.1. An eighth-grade student's link sheet for the procedure of simplifying algebraic expressions

The same class used the link sheet to connect the ideas associated with the concept of variable. The "Diagram/picture/graphic" box again proved useful. At first, most students drew a flow diagram of an equation, which had been the subject of recent lessons. After prompting from the teacher, the students realized that a graph was another way of representing a relationship with variables. This representation had not been used recently.

Not surprisingly, furnishing their own explanations proved difficult for the students, but the exercise provided the teacher with an idea of the students' conceptions of variable. Some students' responses to this section are shown in figure 5.2. Most of the responses were very brief and did not really convey the meaning of variable, which indicated that the students were not able to communicate the meaning verbally. This information prompted the teacher to have a follow-up discussion with the whole class in which he helped the students use their experiences of the previous weeks to talk about what a variable is.

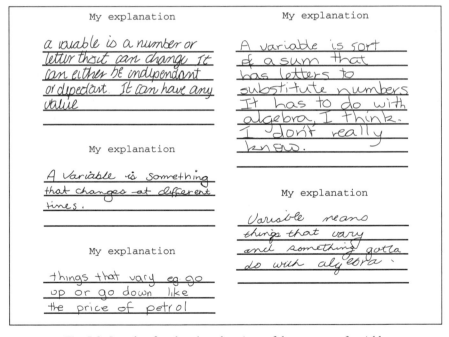

Fig. 5.2. Sample of students' explanations of the concept of variable

A Diagnostic Tool

A different approach was used with a group of preservice teachers in a university methods class. At the commencement of each topic and before any group discussion, the students silently reflected on the idea or procedure, then individually completed a link sheet. Groups of four or five students were formed to discuss the information on their link sheets. During the group discussions, the students were often able to expand their knowledge and understanding of the idea or procedure, and many misconceptions were highlighted and often rectified. In this part of the lesson, the instructor moved among the groups, interacting as necessary. At the conclusion of the group activity, students again briefly reflected on the concept and then added to or modified their link sheets. Used in this way, the link sheet helps students clarify and develop their conceptions. For the instructor, an examination of the students' material indicates the current level of their development and highlights their misconceptions.

Using the link sheet immediately before introducing major ideas yields important information about students' preconceptions and misconceptions. This knowledge forewarns the teacher, who is then able to plan appropriate activities to overcome any problems and to cater to individual needs. For example, one preservice teacher, when completing the link sheet on the idea of decimal number, simply placed a large dot in the "Diagram/picture/graph" box and for her explanation wrote, "A decimal is a number with a decimal point." For this student-teacher, the most important idea concerning decimal numbers was the decimal point. Initially she wrote nothing to indicate that she thought of decimals as representing fractions, nor did her writing suggest any understanding of why the point was used. After participating in a small-group discussion, she added the following sentence to her explanation: "Decimals are fractions." Later she confided to the instructor, "I had never thought of decimals as being fractions. They were simply another type of number you learned in school." The student had obviously expanded her knowledge, but more important the instructor had gained valuable information about the student.

CONCLUSION

The examples given in this paper illustrate ways in which link sheets have been used. Obviously, they can be used beneficially in other ways in a mathematics classroom. The use of the link sheet enhances communication among students, acts as a catalyst for the clarification of students' ideas, and provides the teacher with a valuable tool for "viewing" students' conceptions and misconceptions.

REFERENCES

Borasi, Raffaella, and Barbara Rose. "Journal Writing and Mathematics Instruction." *Educational Studies in Mathematics* 20 (November 1989): 347–65.

Miller, L. Diane. "Writing to Learn Mathematics." *Mathematics Teacher* 84 (October 1991): 516–21.

Rose, Barbara. "Writing and Mathematics: Theory and Practice." In *Writing to Learn Mathematics and Science,* edited by Paul Connolly and Teresa Vilardi, pp. 15–30. New York: Teachers College Press, 1989.

Swing, Susan, and Penelope Peterson. "Elaborative and Integrative Thought Processes in Mathematics Learning." *Journal of Educational Psychology* 80 (March 1988): 54–66.

6

Using Multiple Representations to Communicate: An Algebra Challenge

Leah P. McCoy

Thomas H. Baker

Lisa S. Little

THE traditional symbol-manipulation algebra courses in which students learned to simplify algebraic expressions and solve equations with little connection to real-world applications are no longer sufficient. We need to foster students' exploration of algebraic models in real-world contexts using multiple representations (Glatzer and Lappan 1990; Kaput 1989).

This multitask approach means that students should be able to translate freely among multiple representations: words, tables, equations, graphs. For example, if a student is given a function such as $y = 4x - 2$, he or she should be able to describe problem situations for which the equation would be used. Similarly, given a graph, the student should be able to write "the story of the graph," translating it into words. In traditional courses, we ask students to translate *from* words, but the new emphasis is to have students demonstrate their understanding by also translating *to* words from other representations. Wagner and Kieran (1989) identify problem representation in an algebraic system as an important feature in algebra learning.

Students are better able to "make sense" of a concept when they discuss the mathematics with peers and teachers (Lodholz 1990). Forming the understanding of a concept into words forces metacognitive activity and thus improves thinking. Research studies have reported an increase in mathematical learning as a result of requiring students to share their thinking (Russell and Corwin 1991). Students seek understanding by discussion, including conjecturing, arguing, and justifying. Here is an activity for algebra classes that promotes such learning.

A group of students at work

THE CHALLENGE

Working in cooperative groups, students translate among four representations: words, tables, graphs, and equations. Initially, the groups are given
one of the four representations for a problem and asked to produce the
other three. The groups are encouraged to work together and to discuss
the problem. For example, they may be given a table of x- and y-values
and asked to create a story for these data and then draw the graph and
write the equation. Or they may be given a graph and asked to create a
story, make a table, and write an equation. Students engage in this activity
once or twice a week until they are comfortable translating among the
multiple representations.

Once this background is established, the real challenge begins. Each
group creates an original problem using one of the four representations
(on a given day, everyone may give a word representation or on another
day, an equation). (See figs. 6.1 and 6.2.) Another group is then challenged to represent the problem in the other three modes. At the end of
the session, all the groups present their problems in all representations to
the entire class, using overhead transparencies or large poster boards.
Groups usually do not create exactly the same "exhibit" as the challenging group had in mind, and such a response is a good stimulus for a discussion of multiple correct answers. Graphing calculators or computer
programs can be used for generating graphs. Students need experiences
representing algebra both with and without technology.

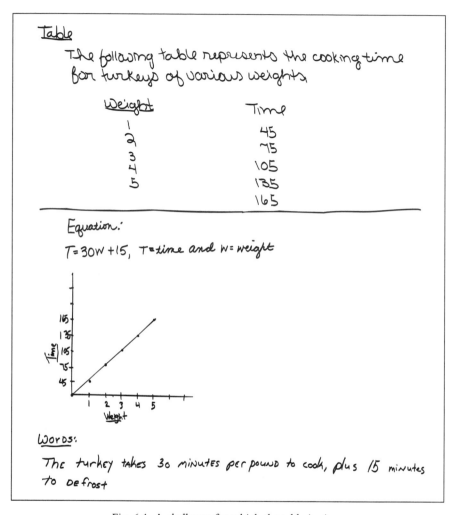

Fig. 6.1. A challenge for which the table is given

An important extension of this activity is to use a graphing calculator to find a regression equation. This variation works well for laboratory activities when the graph of the data is not perfectly linear. The other aspects of the activity remain the same except that the students actually collect the data, which they record in a table. The challenge is to use the graphing calculator to display the data in graphic form, to obtain a regression equation, and then to describe the relationship in words. The following sample laboratory experiments can yield data for this class activity:

- Blow a soap bubble and record its approximate diameter and the number of seconds it takes for the bubble to burst in a controlled environment.

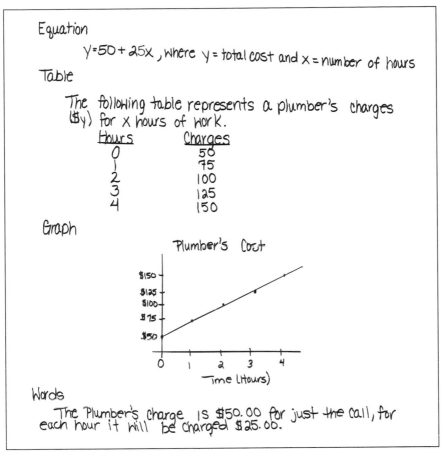

Fig. 6.2. A challenge for which the equation is given

- Set up different numbers of dominoes; record the number of dominoes in each setup and the length of time each takes to fall down.
- Set up a candle and record the elapsed time and the candle's length every five minutes until it is burned.
- Find a cricket outside your house and record the number of times it chirps each minute and the temperature.

CONCLUDING COMMENTS

This algebra challenge provides many experiences that connect algebraic situations to multiple representations. The students are encouraged to employ a multitask approach as they discuss, plan, and create their group exhibit of the four representations of simple functions. This activity gets all students involved in experiencing and communicating about multiple representations. As a result, the algebra classroom becomes a richer learning environment.

BIBLIOGRAPHY

Glatzer, David J., and Glenda Lappan. "Enhancing the Maintenance of Skills." In *Algebra for Everyone,* edited by Edgar L. Edwards, Jr., pp. 34–44. Reston, Va.: National Council of Teachers of Mathematics, 1990.

Kaput, James J. "Linking Representations in the Symbol Systems of Algebra." In *Research Issues in the Learning and Teaching of Algebra,* edited by Sigrid Wagner and Carolyn Kieran, pp. 167–94. Research Agenda for Mathematics Education, vol. 4. Reston, Va.: Lawrence Erlbaum Associates and National Council of Teachers of Mathematics, 1989.

Lodholz, Richard D. "The Transition from Arithmetic to Algebra." In *Algebra for Everyone,* edited by Edgar L. Edwards, Jr., pp. 24–33. Reston, Va.: National Council of Teachers of Mathematics, 1990.

Rosenberg, Jon. Math Connections: Algebra I. [Computer software] Pleasantville, N.Y.: Sunburst/Wings for Learning, 1991.

Russell, Susan Jo, and Rebecca B. Corwin. "Talking Mathematics: 'Going Slow' and 'Letting Go.'" In *Proceedings of the Thirteenth Annual Meeting of the North American Chapter of the International Group for the Psychology of Mathematics Education,* edited by Robert G. Underhill, pp. 175–81. Blacksburg, Va.: Psychology of Mathematics Education, 1991.

Wagner, Sigrid, and Carolyn Kieran. "An Agenda for Research on the Learning and Teaching of Algebra." In *Research Issues in the Learning and Teaching of Algebra,* edited by Sigrid Wagner and Carolyn Kieran, pp. 220–37. Reston, Va.: National Council of Teachers of Mathematics, 1989.

Wah, Anita, and Henri Picciotto. *Algebra: Themes, Tools, Concepts.* Oak Lawn, Ill.: Creative Publications, 1994.

Winter, Mary Jean, and Ronald J. Carlson. *Algebra Experiments I: Exploring Linear Functions.* Palo Alto, Calif.: Dale Seymour Publications, 1993.

7

Algebraic Thinking, Language, and Word Problems

Warren W. Esty

Anne R. Teppo

MANY students progress through four years of high school mathematics and yet emerge at the end without acquiring the ability to solve word problems (Kieran 1992; Simon and Stimpson 1988). This is not simply a missing skill but an indication of a fundamental deficiency in students' abilities to think algebraically. This article examines student performance on word problems as a way to illustrate the critical role that having an appropriate language plays in the student's ability to think and communicate mathematically.

An important aspect of communication is having some way to represent the object or concept under consideration. It is difficult to think about, let alone talk about, entities that cannot be represented by some kind of word, symbol, or picture. "Language is important because by mention of a word parts of a structure can be called up" (van Hiele 1986, p. 86). In mathematics, symbolic language fills a dual role as an instrument of communication and as an instrument of thought by making it possible to represent mathematical concepts, structures, and relationships (Kaput 1989; Esty and Teppo 1994).

ALGEBRAIC AND ARITHMETIC THINKING

Algebraic language is a concise and efficient medium with which to express mathematical thoughts. Any meaningful sentence using algebraic symbols represents a communication about some mathematical object. This article distinguishes two hierarchical levels of thought (arithmetic and algebraic) that affect students' abilities to think about these objects and use algebraic symbols.

Many students, if given an equation, can manipulate the algebraic symbols correctly. However, these same students are unable to set up such an equation given its relationships expressed in the form of a word problem (Kieran 1992). This deficiency is in part related to students' inability to move from arithmetic to algebraic thinking. A conceptual change needs to occur in students' thinking as they move from arithmetic to algebra. The focus of thought must shift from *number* to *operations* on numbers and *relationships* between numbers.

The following problems demonstrate the need to make operations, rather than numbers, the primary objects of consideration in finding solutions to certain types of word problems. The problems also illustrate the distinction between algebraic and arithmetic thinking, and how the use of algebraic language facilitates thinking at a more abstract level.

Problem 1: A circular walk is inscribed in a square block. The area outside the circle and inside the square will be a garden (fig. 7.1). If the side of the square is 100 feet, how many square feet will the garden be?

Problem 2: A circular walk is inscribed in a square block. The area outside the circle and inside the square will be a garden (fig. 7.1). If the area of the garden is 1 400 square feet, how long is the side of the square?

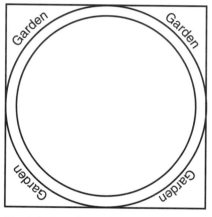

Fig. 7.1. Quiz figure for problems 1 and 2

Both problems describe the same quantitative relationship in words and ask for a numerical answer. Problem 1 asks for an area when a side is given. Problem 2 asks for a side when an area is given.

Each step in problem 1 is a numerical calculation using a well-known formula: 100^2 is the area of the square; $100/2$ is the radius of the circle; $\pi(50^2)$ is the area of the circle; $100^2 - \pi(50^2)$ is the answer. At no stage do we need to use a symbol, such as x, to represent an unknown.

The operations are apparent in problem 1, but they are not the explicit focus of attention. The sequence of operations is the following: (1) *square* the side of the square; (2) separately *divide by 2* to find the radius; (3) *square* the radius; (4) *multiply by* π to find the area of the circle; (5) *subtract* the area of the circle from the area of the square. In problem 1, attention is focused on the numbers—the results of the calculations carried out at each step.

Problem 2 represents a different level of thought. To find the side when the area is given, the student must *represent* this sequence of operations symbolically without actually executing any calculations. Algebraic notation is designed for this job and makes it possible to "build a formula" for the area of *any* garden of that shape (regardless of the side of the square):

(1) $$A(x) = x^2 - \pi(x/2)^2$$

Then, according to the word problem,

(2) $$x^2 - \pi(x/2)^2 = 1\,400.$$

Notice that the remaining steps in the solution manipulate operations (but do not perform calculations):

$$x^2 - \pi(x/2)^2 = 1400$$
$$x^2 - \pi(x^2/4) = 1400$$
$$x^2 - (\pi/4)\, x^2 = 1400$$
$$(1 - \pi/4)\, x^2 = 1400$$
$$x^2 = \frac{1400}{1 - \frac{\pi}{4}}$$
$$x = \sqrt{\frac{1400}{1 - \frac{\pi}{4}}}$$

The solution *process* (but not the solution) would be identical if the area were some other number besides 1400. Numbers are not the focus of attention. This problem illustrates the purpose of algebraic notation, which is to represent operations and order (without actually doing those operations). The critical steps are to represent the relevant operations symbolically and then to manipulate the operations. (The only alternative to using algebraic symbolism is to repeatedly use the operations with specific numbers in a guess-and-check procedure.)

Why are word problems thought to be hard? Is it because the relevant mathematical relationships are difficult to extract from the words? Problems 1 and 2 can be used to investigate that possibility because the relevant mathematical relationships are identical and even expressed in identical words.

Simply using words to describe a problem does not necessarily make it algebraic. If students can do problem 1 but not set up the algebraic equation (2) in problem 2, they are not thinking algebraically. Algebraic thought would be reflected by the use of symbols to help represent the essential conceptual objects (operations and order) of problem 2.

A STUDY

At Montana State University the students enrolled in our precalculus class have all taken three or four years of high school mathematics and intend to take engineering calculus, but their placement scores are too low to allow them to enroll in calculus. That is, these are students who have taken enough algebra for calculus but who, nevertheless, do not perform well at the prerequisite level.

Experience shows that almost all students who enroll in precalculus perform well at problems that require only plugging in to well-known formulas, regardless of the number of steps, but most are unable to do even short algebraic word problems if the relevant formula is not well known. We began to hypothesize that the difficulty these students had with word problems was not in the understanding of the mathematical relationships expressed in the words, but with the underlying algebraic concepts and the use of symbolism required to express these concepts.

A short action-research study was carried out to investigate this hypothesis. On the first day of one term of precalculus, all 137 students in five sections, mostly freshmen, were given either quiz 1 or quiz 2 (below), each containing one arithmetic and one algebraic word problem. One problem was to evaluate the area of a figure given a side, and the other was to solve for a side given an area. The pictures and the layout of the two quizzes were identical. The difference was that the types of problems were switched on the two quizzes. Problems 1 and 4 use arithmetic thought. Even though each requires several steps, each operation is simply a numerical calculation. In contrast, problems 2 and 3 use algebraic thought because the work requires operations to be represented without actually executing them.

Quiz 1

(Problem 1, discussed above.) (Find the area given the side.)

Problem 3: A pen has the pictured shape (fig. 7.2). If the enclosed area is 32 square feet, what is the length of the top side in the picture?

Quiz 2

(Problem 2, discussed above.) (Find the side given the area.)

Problem 4: A pen has the pictured shape (fig. 7.2). If the length of the top side in the picture is 3.5 feet, what is the area of the pen?

The students' responses to these two quizzes were evaluated and categorized according to whether operations were employed appropriately. The students' use, or lack of use, of a variable to express operations was noted in the solutions to the algebraic problems (2 and 3).

The responses to all four questions were categorized according to whether they exhibited the following characteristics: correct operations in the correct order (category A); some correct operations, but not all correct (category B); no correct operations (category C).

For example, in problem 2, those students who wrote an incorrect or incomplete equation or expression that nevertheless contained x^2 were placed in category B because "x^2" represents the necessary operation of squaring. The mere writing of the formula $A = \pi r^2$ in problem 2 or $A = (1/2)bh$ in problem 3 did not classify a response in category A or B if the formula was not also tied to the given problem situation.

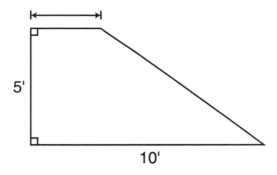

Fig. 7.2. Quiz figure for problems 3 and 4.

In problem 4 (an arithmetic problem), students who found only the area of the left rectangle (obtained by dropping a perpendicular to the base) were placed in category B because some operations were correct. In problem 3, the corresponding algebraic problem, students who set up a correct equation fell in category A. Those who expressed x and $10 - x$ or x and $5x$ were classified in category B if the rest of their answer was incorrect, because at least one correct operation was expressed. The distribution of student responses across all three categories for the four problems is shown in table 7.1.

TABLE 7.1
Numbers and Percentages of Students Using or Representing Operations Correctly: Results of 137 Freshmen Taking Quiz 1 or Quiz 2 the First Day of the Semester

	All Operations Correct	Some Operations Correct	No Operations Correct	Total
Arithmetic thinking (operations used)				
Figure with circle in square problem 1, quiz 1	45 63.4%	12 16.9%	14 19.7%	71 100.0%
Figure with trapezoid problem 4, quiz 2	36 54.6%	21 31.8%	9 13.6%	66 100.0%
Algebraic thinking (operations represented)				
Figure with circle in square problem 2, quiz 2	5 7.6%	9 13.6%	52 78.8%	66 100.0%
Figure with trapezoid problem 3, quiz 1	23 32.4%	10 14.1%	38 53.5%	71 100.0%

The responses to the arithmetic problems 1 and 4 demonstrate that most of the students answering these problems knew what operations to use to find each area. In contrast, the students responding to the similar algebraic problems 2 and 3 did not use those same operations in an algebraic fashion. Far more students demonstrated correct use of operations when only arithmetic thought was required. Remarkably, in each algebraic problem, more than half the students did not express operations algebraically at all.

The circle-in-square problem shows the difference between arithmetic and algebraic performance most dramatically. The fact that 63.4 percent of the students responding to the arithmetic problem used operations correctly to evaluate the area indicates that they understood and could select the appropriate formulas for the mathematical relationships involved in this problem. But only 7.6 percent of those responding to the algebra problem demonstrated that they could correctly express the problem situation corresponding to these relationships when they had to build their own symbolic formula (and all who expressed it correctly solved it correctly, too). Only 21.2 percent of the responses to the algebraic form of the problem contained *any* correct use of algebraic notation at all. All those students who failed to use algebraic notation were not even on the right track.

The categorization of the responses for the trapezoid problem shows similar responses, but the distribution of students across the categories in the table is not so dramatic. However, the reason they are not so dramatic reveals further information about the students' affinity for an arithmetic approach to a solution.

One approach to problem 3 is to drop a perpendicular to the base and create a formula for the area of the rectangle on the left and the triangle on the right: $A(x) = 5x + (1/2)5(10 - x)$. Then set this equal to the given area, 32, and solve for x. Fourteen of the twenty-three students who got this problem right (of the seventy-one who tried) used this algebraic method. Nine used a more numerical approach. By completing the rectangle (fig. 7.3), they were able to deduce that the triangle had area $5(10) - 32 = 18$. Then they solved $A = (1/2)bh = (1/2)5h$ for h and subtracted the number h from 10 to find the desired side. This sequence avoids building a symbolic formula. Rather, students are able to work with known numbers except for the single step, which they solve using the well-known formula for the area of a triangle.

The responses to the set of paired problems with the same mathematical relationships show that many students with a great deal of training in algebra are still far more comfortable identifying and using operations on known numbers than they are expressing these same operations when the numbers are unknown. These responses can be interpreted as an indication of a difference between *being able to use operations* and *having operations as conceptual objects* that can be dealt with without reference to particular numbers. These two types of behavior characterize a difference between arithmetic and algebraic thought.

The high percentages of successful responses for the arithmetic form of each problem indicate that most of the students understood the quantitative situations described in words well enough to select the correct operations in the correct sequence when numbers could be used at each step. But the data also show that many of the students are not able to move beyond an arithmetic level of understanding to think algebraically about those same operations when the number is unknown.

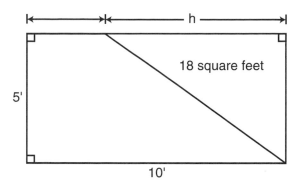

Fig. 7.3. Students' arithmetic approach to problem 3

UNKNOWN AND DUMMY VARIABLES

Students who perceive algebraic symbols as representations of specific numbers and combinations of symbols as directions for manipulating numbers will have difficulty perceiving algebraic objects at the level of operations and order. This section examines how a numerical or algebraic perception of the concept of *variable* affects students' abilities to think algebraically.

Love (1986, p. 49) defines algebraic thinking as

> not merely "giving meaning to symbols," but another level beyond that: concerning itself with those modes of thought that are essentially algebraic—for example, handling the as-yet-unknown, inverting and reversing operations, seeing the general in the particular. Becoming aware of these thought processes and in control of them, is what it means to think algebraically.

In this definition, being able to "handle" the as-yet-unknown really means being able to recognize, express, and manipulate operations. To function at this level, a student's perception of algebraic entities must shift from *number* to a focus on *operations* and *relationships* among numbers. Different kinds of variables are used to express these different kinds of thoughts.

Unknowns are variables (symbols, usually letters) used to represent *particular* numbers in thoughts about numbers. For example, the equation $3x + 2x = 20$ gives information about "x," a number. (The term *variable* may be unfortunate, since some variables, such as unknowns, are not intended to "vary" [Schoenfeld and Arcavi 1988]. In contrast, dummy variables, also known as placeholders, are letters used to hold places where *any* number could be used in thoughts about operations and order (Usiskin 1988; Esty forthcoming). For example, the identity $3x + 2x = 5x$ gives information about the operations of multiplication and addition. This identity gives no information about x, which is merely a placeholder.

The conception of algebra as being *about* operations and order is required in order to appreciate the way that written mathematics employs dummy variables to express methods using identities, theorems, and formulas (Esty 1993; Esty forthcoming). Identities express alternative sequences of operations using dummy variables. How do we subtract a negative number from a positive number? The identity $a - (-b) = a + b$ expresses a method for evaluating $5 - (-3)$. It holds for all values of a and b. The symbols a and b hold places, but not particular values.

Theorems that express the equivalence of equations use dummy variables. The following is an example: $10^a = b$ is equivalent to $a = \log b$. The purpose of a and b in this theorem is to hold places so that the relationship between the operations of exponentiation and taking logarithms can be expressed. The symbol a could be replaced in both places by c or x without affecting the meaning.

Reconsider the algebraic version of the problem with a garden outside a circle in a square (problem 2, fig. 7.1). The key formula $A(x) = x^2 - \pi(x/2)^2$ holds for any side and contains no information about the numerical value of x, the side. The subsequent equation employs x as an unknown, but the numerical information it contains is not relevant until the final, computational, step in the solution process.

The use of letters as dummy variables is a characteristic of algebraic thought. Algebraic concepts and symbolic algebraic language reinforce each other. Without the language, such concepts are difficult to articulate. At the same time, the existence of these concepts creates a need to develop a language appropriate for their study.

CONCLUDING COMMENTS

The students' responses to the action–research study illustrate key differences between arithmetic and algebraic thinking. These differences affect students' abilities to use symbolic notation to think about operations and order. However, the process of moving from arithmetic to algebraic thinking is nontrivial (Sfard and Linchevski 1994). Most precalculus students in the study did not represent operations when it was appropriate to do so. This evidence is particularly striking because it suggests that the students did not grasp the very purpose of the algebraic notation, in spite of using the notation daily in two or more years of high school algebra.

The data from the parallel problem sets also indicate the potential for a serious problem with mismatched instruction. Students who are at a numerical level of thought will not necessarily be prepared to work with algebraic objects of thought at the conceptual level assumed by algebra texts and instructors.

The symbolic language of algebra takes on meaning as students study the ways in which it extends their ability to think about mathematical concepts. By studying word problems, students encounter examples in which algebraic language and the use of dummy variables enable them to think in new, more abstract ways about the conceptual objects of algebra: operations and order.

"Algebra is a symbol system of unparalleled power for communicating quantitative information and relationships" (Fey 1989, p. 207). If students are to be given access to this power, then it is important that we understand and make explicit to students the underlying concepts of the discipline. Symbolic notation is essential for conceptualizing algebra.

REFERENCES

Esty, Warren W. *The Language of Mathematics.* Bozeman, Mont.: Warren Esty, 1993.

——. *Precalculus Concepts.* Bozeman, Mont.: Warren Esty, forthcoming.

Esty, Warren W., and Anne R. Teppo. "A General-Education Course Emphasizing Mathematical Language and Reasoning." *FOCUS on Learning Problems in Mathematics* 16.1 (Winter 1994): 13–35.

Fey, James T. "School Algebra for the Year 2000." In *Research Issues in the Learning and Teaching of Algebra,* edited by Sigrid Wagner and Carolyn Kieran, pp. 199–213. Reston, Va.: National Council of Teachers of Mathematics; Hillsdale, N.J.: Lawrence Erlbaum Associates, 1989.

Kaput, James J. "Linking Representations in the Symbol Systems of Algebra." In *Research Issues in the Learning and Teaching of Algebra,* edited by Sigrid Wagner and Carolyn Kieran, pp. 167–94. Reston, Va.: National Council of Teachers of Mathematics; Hillsdale, N.J.: Lawrence Erlbaum Associates, 1989.

Kieran, Carolyn. "The Learning and Teaching of School Algebra." In *Handbook of Research on Mathematics Teaching and Learning,* edited by Douglas A. Grouws. New York: Macmillan Publishing Co., 1992.

Love, Eric. "What *Is* Algebra?" *Mathematics Teaching* 117 (December 1986): 48–50.

Schoenfeld, Alan H., and Abraham Arcavi. "On the Meaning of Variable." *Mathematics Teacher* 81 (September 1988): 420–27.

Sfard, Anna, and Liora Linchevski. "The Gains and Pitfalls of Reification—the Case of Algebra." *Educational Studies in Mathematics* 26 (1994): 191–228.

Simon, Martin A., and Virginia C. Stimpson. "Developing Algebraic Representation Using Diagrams." In *The Ideas of Algebra, K–12: 1988 Yearbook,* edited by Arthur F. Coxford, pp. 136–41. Reston, Va.: National Council of Teachers of Mathematics, 1988.

Usiskin, Zalman. "Conceptions of School Algebra and Uses of Variables." In *The Ideas of Algebra, K–12: 1988 Yearbook,* edited by Arthur F. Coxford, pp. 8–19. Reston, Va.: National Council of Teachers of Mathematics, 1988.

van Hiele, P. M. *Structure and Insight: The Theory of Mathematics Education.* Orlando, Fla.: Academic Press, 1986.

8

Communicating the Mathematics in Children's Trade Books Using Mathematical Annotations

Pamela A. Halpern

THIS paper is concerned with forming closer links between communicating mathematics through the written word and communicating mathematics through written symbols by adding explicit mathematical annotations to children's trade books that contain mathematical concepts. Trade books are written for the purpose of entertaining, not for teaching reading or teaching mathematics. Those containing mathematical themes, however, are being used in mathematics classrooms. Most of these books include no specific mathematical representations. A study conducted to examine the effects on children and adults of adding explicit mathematical annotations to children's trade books will be discussed. Questions include whether such annotations enhance or detract from the trade book and whether the explicit mathematical annotations communicate the mathematical concepts of the story more clearly.

The intent of enhancing children's trade books with explicit mathematical annotations is to highlight the mathematics found in literature and to give children another means by which to construct mathematical knowledge.

ENHANCING CHILDREN'S TRADE BOOKS WITH EXPLICIT MATHEMATICAL REPRESENTATIONS

Children's trade books can be instrumental in communicating connections between mathematics and the real-life situations from which the mathematics naturally springs. Children's trade books containing mathematical concepts have become quite popular for making connections and for motivating students.

Explicit mathematical annotation is not often found in children's trade books. Yet it would present an opportunity to strengthen the communication of concepts and of applications of mathematics.

The first question is whether the explicit mathematical annotation is negatively distracting to the readers, both children and adults. Does this annotation add to, or detract from, the reader's enjoyment of the trade book? The study reported here explored this question.

The second question is whether children's trade books enhanced with explicit mathematical annotations communicate the mathematical content more effectively than the unenhanced books. An analysis of the data gathered in the study sheds some light on this question.

The mathematics in three children's trade books that contain mathematical concepts was enhanced by the addition of explicit mathematical representations in the form of number sentences and graphs. The addition was as natural as possible. Children were then read either the original or the enhanced version of one of the books and asked what they thought the book was about. They were then read the other version and again were asked what the book was about. They then chose the version of the book they would prefer to take home. The books chosen for the study are *I'll Teach My Dog 100 Words* (Frith 1973), *Alexander, Who Used to Be Rich Last Sunday* (Viorst 1978), and *The Doorbell Rang* (Hutchins 1986).

I'll Teach My Dog 100 Words is about a youngster who plans all the things he will teach his puppy. Written in rhyme, it lists in context the 100 words the child will teach the dog. Periodically throughout the book, a clipboard appears in the margin, along with a hand and pencil. On the clipboard is a single number representing the tally of the number of words the child will teach the dog. The mathematical enhancements to this book include an addition number sentence on the clipboards and a bar graph in each outside margin of the double-page spread. The graph is shaded to indicate the number of words the puppy has been taught at different points in the book.

Alexander, Who Used to Be Rich Last Sunday is about a young boy who has a hard time hanging on to his money. The story tells how the dollar his grandparents gave him "only last Sunday" was "spent." The enhancement in this book consists of subtraction sentences that appear on each page on which Alexander's money supply dwindles. Eventually, Alexander is left with a few memories and mementos and the bus tokens he started with.

In *The Doorbell Rang,* the tension mounts as every time the doorbell rings more and more friends come to visit and share the cookies that Mom has just taken out of the oven for Sam and Victoria. Grandma saves the day when at the end of the book, she appears at the door with an enormous tray of cookies. The enhancement is a division equation that represents the equal sharing of Mom's original dozen cookies among the children present at different times in the story.

THE STUDY

The sample population for the study consisted of 252 subjects (209 children and 43 adults). All the children were students in grades 1 through 3 in one of two elementary schools in an upper-middle-class suburb. The adult sample consisted of 14 parents of students participating in the study, 9 teachers, 4 content-area specialists, 5 teacher educators, and 11 college students in preservice teaching programs. In educational achievement, they ranged from high school graduates to those with postgraduate degrees. Although the participants were not asked to discuss their attitudes about mathematics, a strong trend developed. The participants who specialized most in either mathematics or reading and those who volunteered a preference for either mathematics or reading tended to prefer the original versions of the trade books.

Do Annotations Detract from Readers' Enjoyment?

Given a choice between the original and the enhanced versions of the three trade books, 82.1 percent of the entire sample—children and adults—favored the enhanced version, whereas only 17.9 percent favored the original version. Among the children, 81.8 percent favored the enhanced version and 18.2 percent favored the original. The enhanced version was preferred by 83.7 percent of the adults; 16.3 percent preferred the original. Thus the enhanced version was preferred by a significant majority of the subjects. In answer to the first question, enhancing children's trade books with explicit mathematical annotations appears not to have negatively distracted a majority of the readers. Although the subjects were not directly questioned about either their comprehension of the story or about distractions the enhancements may have caused, they were asked to relate what they thought the story was about and to state the reasons why they had made their choice of the two versions. Among the 17.9 percent of the subjects who preferred the original version, none mentioned the enhancements as detracting from the story. A majority of the readers—both children and adults—who preferred the mathematically enhanced versions noted that the enhancements added to their understanding of the story.

The subjects were asked not only to give their preferences but to state reasons for them. In the qualitative analysis of these data, the following reasons were among those recorded for preferring the enhanced versions of the trade books:

- The mathematical annotations made reading the book more enjoyable and even fun.
- The mathematical annotations made the book "easier to understand" and communicated the mathematics more clearly.

The following observations were made about those who preferred the enhanced version:

- Many of the readers became engaged in the mathematics of the enhanced story by verbally predicting what would happen next and by verbally offering alternative mathematical representations of the situation found in the literature.
- The adults saw the mathematical enhancements as encouraging those students who prefer reading over mathematics—or conversely, mathematics over reading—to see the relationship between the two modes of communication and to encourage participation in the less favored mode.

Do Annotations Communicate Mathematics More Effectively?

The second question involves the role of explicit mathematical annotations in communicating the mathematics in children's trade books. The reasons given by subjects support the conjecture that explicit mathematical annotations in children's trade books facilitate the communication of the mathematical concepts in the story.

No significant difference was found in the subjects' perception of the storyline in the enhanced and original versions of any of the three books. The addition of explicit mathematical annotations did not cause a significant number of readers to change their original opinions of what any book was about.

Among the books that are most appropriate for enhancement with mathematical annotations are those that have a mathematical theme throughout the story. The books chosen for the study lend themselves to straightforward algorithmic treatments of mathematical concepts that continue throughout the stories. They can be considered predictable books or pattern books in that it is clear to the reader what will happen next (predictable) and key words or phrases are repeated throughout the text (pattern). Not all books containing mathematical concepts fit these categories because the mathematics may not continue throughout the story or be an integral part of the story. These books or stories can serve as catalysts to mathematical conversation and discovery. For example, *The Paperbag Princess* (Munsch 1980) is a charming story in which a dragon flies around the world in ten seconds and then repeats the performance in double the time, twenty seconds. This incident in the story can be used to stimulate investigations of speed calculation, large numbers, small numbers, and unit conversion.

CLASSROOM IMPLICATIONS

Much mathematics occurs in children's literature—some good, some bad; some correct, some incorrect—but it has been observed that most of it is ignored because the books that contain mathematics do not communicate it explicitly in the manner to which most people are accustomed—through number sentences, graphs, diagrams, and algorithms.

Many publications suggest that children's literature be used in conjunction with mathematical topics in the elementary grades (Welchman-Tischler 1992; Thiessen and Matthias 1992; Whitin and Wilde 1992). These publications do not often advise enhancing the literature with explicit mathematical annotation. Teachers should not be afraid to introduce explicit mathematical annotations and vocabulary with this literature. Adding the annotations or having the students add the annotations can serve as a lens through which mathematics can be seen as more than a rigid, rigorous topic that is composed of symbols and rules and that values accuracy and speed. There is a danger in overenhancing the literature. A delicate balance must be kept so as not to destroy the literature as a total work. If the enhancements become overbearing, they could have a negative effect on the reader.

Directions for Enhancing a Book

The following are suggestions for enhancing children's trade books:

- Select an appropriate book. Many books list children's literature that can be used for mathematical learning. They include excellent resources, and most list titles by grade level and topic (Thiessen and Matthias 1992; Welchman-Tischler 1992; Whitin and Wilde 1992).

- After choosing a book, read it through for sheer enjoyment before you begin dissecting it for mathematical content.

- Decide on enhancements that would highlight the mathematics without detracting from the story.

- Make the appropriate enhancements in the book. Removable adhesive paper works well for enhancements that are not meant to be permanent.

Student Activity

Children are encouraged to annotate trade books themselves. For follow-up and extensions, have the students work first as a class, next in small groups, and then individually on enhancing children's trade books. This activity has proved to be very popular with students. They become very creative in constructing representations of the mathematics that they find. Rather than randomly place annotations on the page, students have used "think" balloons, they have added physical representations and such manipulatives as coins and paper money, and they have designed and included manipulative materials to complement the mathematics in the story.

CONCLUDING COMMENTS

Children's trade books that are enhanced with appropriate, relevant, explicit mathematical annotations that are in concert with the story actually seem to be preferred by both children and adults over the same books with no explicit mathematical annotations. The enhanced versions also seem to communicate the mathematics in the story more clearly. When mathematical annotations are added to children's trade books, the written word and the representations can be used together to communicate mathematics more effectively.

REFERENCES

Frith, Michael. *I'll Teach My Dog 100 Words.* New York: Random House, 1973.

Hutchins, Pat. *The Doorbell Rang.* New York: Greenwillow Books, 1986.

Munsch, Robert N. *The Paperbag Princess.* Toronto: Annick Press, 1980.

Thiessen, Diane, and Margaret Matthias, eds. *The Wonderful World of Mathematics: A Critically Annotated List of Children's Books in Mathematics.* Reston, Va.: National Council of Teachers of Mathematics, 1992.

Viorst, Judith. *Alexander, Who Used to Be Rich Last Sunday.* New York: Atheneum Publishers, 1978.

Welchman-Tischler, Rosamond. *How to Use Children's Literature to Teach Mathematics.* Reston, Va.: National Council of Teachers of Mathematics, 1992.

Whitin, David J., and Sandra Wilde. *Read Any Good Math Lately?* Portsmouth, N.H.: Heinemann Educational Books, 1992.

9

Fostering Metaphorical Thinking through Children's Literature

David J. Whitin

Phyllis E. Whitin

As a university-level–elementary-level teaching team, we have become intrigued by children's use of metaphors to express their mathematical understandings. Metaphors are a natural avenue for children to invent their own language and connect it to the world of mathematical ideas. Early in the school year we noticed fourth-grade children using metaphors quite spontaneously to describe various learning experiences. When Danny circled multiples of 4 and 8 on a hundred chart, he noticed that sometimes multiples of both 4 and 8 landed on a common number (8, 16, 24, etc.). He remarked, "Four is just like eight's little sister. She follows him wherever he goes." His analogy helped the class look more closely at common multiples and how they related to counting-on numbers. This natural occurrence of metaphor early in the year convinced us that we ought to encourage its use. We knew that historically, mathematical vocabulary had been controlled by the textbook and that little attention had been paid to the language of children. We were beginning to see the potential that metaphors hold for restoring visual imagery to mathematical ideas. Since children (and all learners) naturally frame their world in metaphors, we wanted to be sure that the children in our classroom did not leave their language and their world outside the schoolhouse door. Instead, we wanted to draw out their language as a legitimate and valued part of classroom life.

SHARING A PIECE OF CHILDREN'S LITERATURE

One of the ways in which we fostered the use of metaphors was to encourage their use as a response to mathematical stories. We saw children's literature as a powerful avenue for engaging learners in productive conversations and investigations in mathematics (Whitin and Wilde 1995).

Since stories provide a meaningful context for mathematical ideas, we saw them as a natural way for children to tie abstract symbols to their personal world. One of the stories we chose to read to the children was *Two of Everything* (Hong 1993). We thought the story would invite the children to explore a geometric sequence and to create metaphors to describe its exponential growth.

In this story a poor farmer, Mr. Hacktack, digs up an ancient pot from his field. He drags it home, tossing into it his purse with his last five gold coins for safekeeping. After Mr. Hacktack arrives home, Mrs. Hacktack discovers that the pot now contains two identical purses with five coins each. The delighted couple decides to amass a huge pile of gold coins, and Mr. Hacktack heads to the market. Laden with packages on his return, he kicks open the door. It hits Mrs. Hacktack, who falls into the pot, and the troubles begin. The story ends happily with two Mr. and Mrs. Hacktacks living prosperously side by side.

As we read the story aloud, we were intrigued by the children's spontaneous reactions. Some comments reflected a problem-posing stance by inviting a variation of the story. For example, Eric wondered, "What else could you put in the pot?" Tony predicted the numerical sequence when he chimed in, "Ten gold coins!" It was Shunta's comment, however, that helped us reflect on the power of metaphorical thinking. When the second Mrs. Hacktack emerged from the pot, Shunta muttered, "Troubles are beginning to double!" Until this point in the story, only tangible things had doubled, but Shunta realized that intangible ones can multiply as well. Her insight bridged the gap between the concrete aspect of the story and a more abstract aspect. Herein lies the essence of metaphorical thinking: linking the world of sensation to the world of thought through mental imagery (Whitin 1996).

INVESTIGATING THE NUMERICAL SEQUENCE

After completing the story, we asked the children to pick something that would be interesting to double and to continue the sequence with calculators. They discussed their ideas as they worked and in this way continued to build on one another's discoveries. They were amazed that the numbers soon became quite large. Eric remarked, "When I started with 1, I thought it would take a long time, but I got past a million. It went really fast." Shannon wrote in her journal, "I started to get all excited when I got up to the thousands because it made me think of what if you had that much money, especially one million!" Other children noted some patterns. Ashley wrote, "All the end numbers go in a pattern and are even. 2, 4, 8, 6. Keeps on going." With the exception of 1, all the numbers were even, and the last digits did follow that 2-4-8-6 pattern: 1, 2, 4, 8, 16, 32, 64, 128, 256, …. Rett noticed that the first three numbers in the sequence (excluding 1) were single digits (2, 4, 8), the next three were double digits (16, 32, 64), and the next three were three digits (128, 256, 512). Billy saw that at one point the sequence increased by 4 (4 to 8), but that later on, "you skip by the thousands." His comment helped to describe a distinguishing feature of a geometric progression, namely, that it does not increase by a constant added amount. Chris, who decided to start his sequence at 3 (3, 6, 12, 24, 48), remarked, "Doubling by 3s is a lot different than counting by 3s—after you get to the thousands especially." Chris was noting the difference between an arithmetic and a geometric progression.

After the children shared these observations, we wanted to encourage some metaphorical descriptions of this sequence. We therefore asked the children, "What do these numbers remind you of? What picture do you have in your mind as you see numbers like these getting larger and larger?" Some children looked at their increasing column of numbers as a handy reference for their metaphors. Ashley said, "The numbers look like a tower going down," and Andrew wrote, "It looks like Idaho." (See fig. 9.1.) Andrew made a more abstract comparison when he continued, "It reminds me of an avalanche because the numbers are snow fall[ing] down a mountain." Other children created metaphors that were similarly tied to familiar situations, such as reproductive cycles or food. Rhiannon wrote about cats and kittens; Stephanie, pet birds and eggs; and Chris, "plants growing from seeds." Jonathan wrote that the sequence "reminds me of a cookie growing in the oven." Ashley related it to a local doughnut shop that is open twenty-four hours a day: "You can see the doughnuts through this window, and they keep on going, and they put them in boxes, and the boxes get bigger and bigger and bigger." William made the analogy to birds at our classroom feeders: "It's like one bird coming to the feeder first, and then it flies off and tells two of its friends. And they come, and they tell more of their friends, and then four come, and it keeps going."

Although some of these metaphors, such as the doughnut shop, did not technically represent a geometric sequence, we nonetheless considered it important to accept all these metaphorical connections. All the children were finding examples of collections of things that increased over time, and

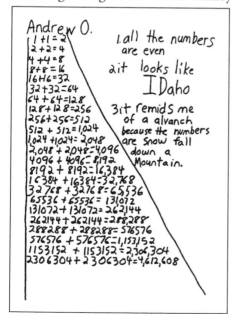

Fig. 9.1. The metaphors of Idaho and the avalanche

that was the important point during this part of our brainstorming. When we discussed these metaphors with the children the next day, we noted that some of these examples, such as an avalanche or a population, show a dramatic increase in size, whereas others double only once (e.g., most cookies double in size while baking) and still others, such as boxes of doughnuts, increase in number at a more regular rate. These personally meaningful images enabled us to distinguish geometric from arithmetic progressions.

EXPANDING ON OUR METAPHORS

We were particularly intrigued with Andrew's avalanche analogy; it captured the nature of a geometric progression through both sight and sound. We began the next class by displaying Andrew's avalanche and asking the children to interpret it. Danielle said, "An avalanche starts off... it pours down. The avalanche starts kind of small, and then starts getting bigger and bigger and bigger." Jenny, remembering the limitations of the calculator from the previous day, remarked, "An avalanche gets bigger and bigger until it hits a tree or something and finally stops. It can't hold any more. Like on the calculator we ran out of numbers."

After the children shared other metaphors from the previous day and brainstormed new ones, we asked them to write and draw these ideas on paper. Shawn and Jonathan, inspired by Andrew, drew snowballs, but Andrew abandoned the avalanche idea in favor of an auditory image. Always fascinated with trucks, he sketched one with the words *far away, close, and near* above it. He labeled these categories with groups of numbers from the geometric sequence: 2, 4, 8 by "far away"; 16, 32, 64, 128, 256 by "close"; and numbers beginning with 512 by "near." He wanted to compare the size of the numbers to the sound of an approaching truck. Smaller numbers conveyed the faint sound of the truck in the distance. As the truck rumbled closer, the numbers became larger. Finally, the largest numbers represented the loudest sound as the truck whizzed past the nearby listener.

Perhaps the idea of a mountain inspired William; he thought of the Eiffel Tower (see fig. 9.2), with the small numbers in the spire and increasingly larger numbers toward the base, where the mass of the structure was reflected in its area, volume, and number of beams. Although William did not actually calculate the measurements of the tower using these concepts, he did show through his drawing that as the sweeping lines of the tower continued to widen, the numbers of the progression continued to increase.

Amanda described the ever increasing problem of pollution when she sketched the ocean filled with litter. She wrote, "This is a picture of a man polluting the ocean and it reminds me of when you start to pollute and it gets bigger and bigger." Rett also thought of an ocean, but in his picture he labeled the shallow parts with small numbers and the ocean depths with large numbers. He wrote, "It reminded me of an ocean because it gets deeper and deeper and the numbers get larger."

Paris

William

I am showing the tower as the numbers grow the larger the tower.

This is the great tower in Paris France shown as a model if it was made out of numbers.

Fig. 9.2. The metaphor of the Eiffel Tower

After the children share their metaphors with one another, it is important to analyze their images in more detail so that they can begin to note the salient differences between an arithmetic and a geometric progression. For example, it might be helpful to begin with an example that demonstrates a constant ratio of growth, such as the spreading of news that William referred to. Children can understand the situation of spreading a piece of news (or a secret!) by describing it this way: one person tells two people, and each of those two people tell two people (now four more people know), and each of those four people tell two people (now eight people know), and so on. The examples of plant and animal reproduction that the children discussed could be examined by looking at cell division, the growth of bacteria, or the branches of a family tree. To highlight the difference between an arithmetic and a geometric progression even further, the children might add by a constant of 2 on the calculator. They would see that the constant difference between a given term and its predecessor does not gain the momentum that the earlier progression did (when the screen was filled with numbers after only twenty-three doublings). These particular discussions and experiences would be helpful to them in looking more critically at their own metaphors: some cookies double once in size, but that is all; dumping garbage in the ocean is indeed an environmental problem, but if the dumping were a geometric progression, the ocean would be filled to capacity in a short time. Thus, the children's metaphors enable them to look more closely at these two numerical progressions as well as analyze more critically the nature of the images.

CONCLUDING COMMENTS

As we look back on this classroom experience, we see important benefits in promoting metaphorical thinking. Metaphors allow children to make personal connections to mathematical ideas. When children frame the concepts of mathematics in their own language, they develop ownership of these ideas. Metaphors also invite everyone into mathematical conversations. Since there is no one-to-one correspondence between a mathematical idea and its visual equivalent, all learners can draw on their personal backgrounds and experiences to offer unique comparisons.

This last point particularly crystallized for us the value of metaphorical thinking. The children were helping us see that metaphors are generalizable because learners of all ages carry these images with them into new situations. As the children developed this rich pool of metaphors to describe a geometric progression, they were also developing a more enlightened eye for describing other learning experiences. Metaphorical thinking in our classrooms can open new and exciting avenues for exploring mathematical ideas.

REFERENCES

Hong, Lily. *Two of Everything*. Morton Grove, Ill.: Albert Whitman & Co., 1993.

Whitin, David, and Sandra Wilde. *It's the Story That Counts: More Children's Books for Mathematical Learning, K–6*. Portsmouth, N.H.: Heinemann Educational Books, 1995.

Whitin, Phyllis. *Sketching Stories, Stretching Minds: Responding Visually to Literature*. Portsmouth, N.H.: Heinemann Educational Books, 1996.

10

Using Reading to Construct Mathematical Meaning

Marjorie Siegel

Raffaella Borasi

Judith M. Fonzi

Lisa Grasso Sanridge

Constance Smith

T<small>HE</small> question of how to foster the kind of thinking that takes students beyond "performing" to sense-making and inquiry lies at the heart of the challenge facing mathematics teachers today. In this article, we wish to show how one form of communication—reading—can find a central place in mathematics instruction so as to engage students actively in meaningful learning.

The examples of "reading to learn mathematics" that form the centerpiece of this article are drawn from a collaborative research project designed by an interdisciplinary team composed of two mathematics education researchers, a reading education researcher, and a group of secondary mathematics teachers. The learning experiences developed within this project were not only consistent with the instructional goals articulated in the NCTM *Standards* documents (1989, 1991) but also aimed at helping students gain a better understanding and appreciation of the nature of mathematics as a discipline. Although this goal is neglected in most mathematics curricula even today, it was a priority in this project, given that students often hold beliefs about mathematics that can negatively affect their attitudes as well as their performance in school mathematics (e.g., Borasi 1990, 1992).

In order to develop reading experiences that supported these goals, however, the popular notion of reading as a set of skills for extracting information from text had to be challenged and expanded. Our work is therefore based on a transactional theory of reading, one in which the reader actively shapes and is shaped by the text (Rosenblatt 1978). This means that readers do not simply *take* meaning from text but use their knowledge, interests, values, and feelings to *generate* meaning. Making sense of a text thus requires that the reader act on it and make it her or his own. When considered in relation to the instructional goals we were trying to support, this alternative view of reading implies a broader definition of *what kinds of texts* could serve as suitable reading material for the mathematics classroom, *how* these texts could be read, and *why* they are read (Borasi and Siegel 1994).

The two vignettes that follow offer concrete illustrations of what a transactional perspective on reading looks like in a mathematics classroom as well as a glimpse of the variety of reading strategies, texts, and purposes for reading mathematics that teachers can consider for their students. The first vignette reports an instructional episode in which learning about the nature of mathematics and learning new reading strategies were the explicit goals of the activity. The second vignette illustrates various uses of reading within the context of a unit on geometry.

Illustrations of "Reading to Learn Mathematics"

Vignette 1: Using a Transactional Reading Strategy to Learn How to Make Sense of a Difficult Text on the Nature of Mathematics

This first vignette is drawn from a course called Math Connections, designed with the goal of helping students rethink their conceptions of mathematics. This course was offered by Judith M. Fonzi to students in grades 9–11 at the School without Walls, an alternative urban high school in Rochester, New York. This reading experience occurred in the second half of this semester-long course in which students read several texts about the nature of mathematics using various transactional reading strategies introduced by the instructor. The purpose of these activities was to teach students ways to become more active and engaged in sense-making, especially when reading difficult texts, and at the same time to challenge the students' views of mathematics as a discipline. These strategies also offered students a means for sharing with their peers what they had gained from the reading, a weakness they had identified when reflecting on a group project that they had previously completed.

The first transactional reading strategy introduced was the "say something" strategy (Harste and Short 1988). "Say something" is based on the assumption that making sense of text is a social event in which readers talk their way through a piece, sharing their responses, questions, confusions, and insights with partners as they go. The strategy thus demonstrates that comprehending a text is not automatic but involves interactions with both the text and other readers.

To give the students an idea of what the "say something" strategy was all about, the instructor invited the students to try out the strategy by silently reading a short paragraph on mathematical connections and sharing their immediate responses with the class. These responses were recorded on newsprint for all to see, which allowed the instructor to point out the way students had capitalized on one another's questions and comments to make sense of the text. The instructor then asked the students first to choose and then to read with a partner using the "say something" strategy one of two short articles, "Mathematics in the Marketplace" and "Mathematics and War," both from *The Mathematical Experience* (Davis and Hersh 1981), a book that explores the nature of mathematics, including its social, historical, and political dimensions.

Shellie and Van's "say something" experience was especially interesting because they worked out an interpretation of the article on mathematics and war by taking on different roles and in the process generated an analogy that connected their world to the text. What follows is an exchange that took place early in their discussion. The students had finished a paragraph describing Napoleon's use of mathematics and had just started reading one describing the use of mathematics during World War II. Their interaction illustrates well the complementary roles the students spontaneously assumed.

Van: Now they're just showing what math is used for, like during the world war times. [Shellie doesn't respond.] Shellie?

Shellie: I'm just trying to put it all together. Why are they all of a sudden bringing in World War II?

Van: Because they went from this dude [referring to Archimedes] to Napoleon. And then they're going in order of history—how mathematics was built into things. You know—like Hieron—he made a puny catapult and now they're making atomic bombs.

Shellie: A catapult. What's that?

Van: Remember—okay [she demonstrates her idea with paper]—it throws things and they like cut the rope and it ...

Shellie: In the Smurfs?

Van: Yeah. When they threw their smurfberries at Gargamel or whatever? Okay?

They continued on in this manner—Shellie constantly raising questions, not willing to remain confused, and Van helping her make sense of the material by explaining, summarizing, and relating the text to experiences a little closer to home.

The last paragraph of the piece that they read suggests that World War I was the chemists' war, World War II was the physicists' war, and that World War III might be the mathematicians' war. When they stopped to discuss this paragraph, Shellie began by saying she wanted to know why the authors said that World War I was the chemists' war. When Van said she didn't know, Shellie jumped in with a hypothesis that chemicals were used in World War I, drawing on something she'd learned in another class. Then they moved on to the next question about why World War II was the physicists' war. After more discussion, Van said, "Let me try to put this in cars. World War I—your first car is a Chevette." Shellie started to make fun of this but Van ignored her and continued to explain the increased dependence of modern warfare on mathematics in terms of the improvements their friend had made in his car over time. When she had finished elaborating on the analogy, Shellie said, "That was a good example, Van. That was really good.... I understood it.... We should tell someone about that." Van agreed, "Yea, we compared our math to Joe's car. I knew he was good for something in school."

As a follow-up to the "say something" experience, each pair was then asked to create a "sketch" that showed what they had learned from the article. This strategy, called "sketch to stretch" (Siegel 1984), invites students to use a different communication system (e.g., drawing) to interpret a text as a way of helping make new connections between the text and their knowledge and experiences. Shellie and Van completed their joint sketch (fig. 10.1) for homework and shared it with the entire class the next day.

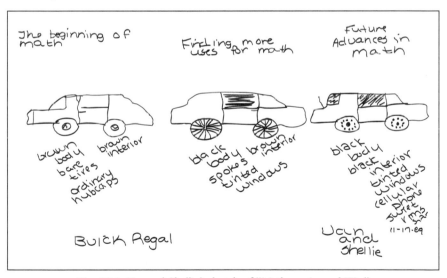

Fig. 10.1. Van and Shellie's sketch of "Mathematics and War"

After Shellie explained the car example, Van related it to the article, saying, "We took it as how the weapons improved through history. They started out with the catapult and now they are into bombs and stuff. And the first picture is like the beginning of math and how math was first discovered to be of use.... And the second picture was when they were finding uses for math for more things in life.... And now [the third picture] is like for future math and now math is used for a lot more things.... And it's almost like a question, too, because there's like more things that he could do to his car and there's more ways that math can be used." What is so intriguing about their sketch is that creating a visual representation of their analogy allowed them to develop the idea that the future of mathematics was like a question, something they had not discussed during the "say something" experience. Van's explanation thus shows how drawing helped them consolidate their analogy and use it as a springboard for generating new insights into the nature of mathematics.

This vignette has provided a clear picture of how some specific transactional reading strategies (such as "say something" and "sketch to stretch") can support students' comprehension of a difficult text dealing with the nature of mathematics and can thus help them reexamine their views of this discipline. Notice how this reading experience helped these students develop a better sense of mathematics as a "growing" discipline with connections to real-life events and experiences. Research (e.g., Borasi 1990) has shown that most students are unaware of these aspects of mathematics and this contributes to their negative perception of mathematics and their sense that studying mathematics has limited usefulness for them.

Vignette 2: Using an Essay on Geometry as a Springboard for Introducing Technical Topics and Engaging in Problem Solving

The second vignette reports on a reading experience that Lisa Grasso Sanridge developed at the beginning of a geometry unit in her eighth-grade middle school mathematics class. Sanridge's middle school is located in a rural community near Rochester, New York. This unit was designed to introduce the students to all the basic geometry concepts usually covered in the eighth-grade curriculum while at the same time helping them to understand better how geometry came about and how it could be used in real life.

The teacher located an interesting and rather accessible essay from *The Whole Craft of Numbers* (Campbell 1976) that discusses how geometry originally developed from people's observations and their need to solve problems in their everyday lives. The essay starts with some speculations about how so-called primitive people might have recognized simple shapes in the world around them and then identified and used some properties of those shapes in simple applications such as constructing dwellings. The essay then discusses several alternative methods for computing distance in a situation where direct measurement is problematic (between two points across a lake). The methods presented to solve this problem all rely on mathematical properties already known to the ancient Egyptians and Babylonians, yet they involve some clever uses of mathematical notions, such as parallelism and the congruence of triangles. Over the course of three class periods, the teacher carefully planned activities around this reading so as to help the students make sense of this complex text and learn something new about geometric figures and, especially, about the congruence of triangles.

The teacher introduced the first part of the text (consisting of about two pages on the beginnings of geometry and concluding with the author's statement of the lake problem) as a reading that would contribute to the class's ongoing discussion about the origin and uses of geometry. She also announced that to help them understand this text, they would read it in a special way—using the "say something" strategy.

To illustrate what this "say something" strategy involved, the teacher demonstrated it on the first few paragraphs with the help of a colleague, inviting the students to join in when they felt like it—as they soon did very spontaneously! The following excerpt from this "class say something" activity illustrates how this reading strategy indeed invited the students to make sense of (and even go beyond!) some of the mathematical concepts discussed in this text, thus contributing to their learning of geometry (Campbell 1976):

> It is also reasonable to assume that the herdsman developed the notion of "straight." Trees grow straight up; rocks fall straight down. Arrows are best made of straight branches. When crossing a meadow or valley the straight path or line is the one which seems to take the least time to travel.

After reading this paragraph silently, several students started talking together, and one of them mentioned "bees," prompting the teacher to ask for clarification:

Teacher: Why bees?

Karrie: They make a beeline towards the flower.

Though several students at first laughed at this explanation, this observation started a worthwhile discussion about what makes a shortest path:

Janet: But when you go long distance, it is no more necessarily true that the straight is also the shortest. If you take a plane or a ship, it starts to make an arc.

Adam: Not if you are on a flat surface!

Janet: Yeah, but if you make a trip on the earth, which is a circle ... so it's shorter to make a round.

Nickolas: The earth is a circle, or is it a sphere?

Susan: So they were not exactly right.

Teacher: My thought is they're herdsmen; you mentioned traveling how?

Janet: I mean, *now* it's not true, because ... [we use] planes.

Kay: Aha!

Craig: But it is still shorter to go straight ...

Janet: And the earth *is* round, Craig!

It is interesting here to note how sharing their responses allowed the students to go beyond the text itself and begin to develop a more sophisticated understanding of the notion of distance as well as straight line. Janet's observation about the path a plane or ship takes generated a lively discussion that led all the students to question what would make the shorter path between two points and helped them realize that, depending on the surface considered, this may not always be a straight line.

After the class read together a few paragraphs in this fashion, the teacher asked the students to share briefly their impressions of this way of reading, and then directed them to break into pairs and continue reading the text "saying something" to their partner for the remaining five minutes of the lesson. For homework, she asked the students to finish the reading on their own and then to respond in writing to the following questions: (1) See if you can come up with at least two ways to measure the distance across the lake; (2) What new things have you learned about geometry?

The next class period began with the students' sharing what they had done for homework. For each of the two questions, volunteers read their responses as the teacher recorded their contributions on newsprint to have a record to which the class could refer as needed. It is important to note how the process of recording students' ideas on newsprint was itself a generative reading event, since students frequently clarified, elaborated, and revised their ideas in response to the questions raised by the teacher and their peers. A seemingly minor activity thus became a significant learning opportunity in which students' original individual texts were transformed into a collective text through a dialogue that involved the negotiation of meaning.

The following excerpts from the students' responses to the first question (i.e., asking for methods to measure the distance between two points A and B across the lake) are a good illustration of the variety and richness of the students' ideas:

> *I could wait 'til the lake freezes, then take a tape measure across it; I could swim across the lake using even strokes, then use the number of strokes as a measuring term.*
>
> —Tim

> *Tie a rope at one side of the lake (point A) then swim across the lake with a rope and cut it at the edge of the lake, then measure the rope; go in a boat and do the same thing.*
>
> —Monica

> *If you extend a line straight out from point A and B and then measure the distance between.*
>
> —Debra

The teacher had intended for this collection of ideas to spark students' interest in the problem of measuring the distance between two points with an obstacle in between and to supply them with something they could rely on when approaching the second—and more technical—part of the reading, where the author explains several mathematically based solutions to this problem. Indeed, as they began to read this new text in class (first "saying something" as a whole class and then in pairs), the students were pleasantly surprised and intrigued by the fact that some of the methods *they* had suggested were mentioned by the author, as well as some new ones. They were also pleased that *they* had come up with some methods the author had not considered! Once again, since the students did not have enough time to conclude the reading in class, they were asked to complete it at home and respond in writing to the questions: (1) What method would you pick to measure the distance from A to B and give your reasons WHY! List advantages. List disadvantages. (2) Will the method mentioned on page 3 work? Explain your answer. Is there another way to make a triangle?

As this second question implies, the teacher had expected that most students would find it difficult to understand the last method described in the reading. This method, based on triangulation, consisted of choosing a third point C (anywhere, provided that AC and BC could be measured easily) and then reproducing the triangle ABC elsewhere (choosing this new location so that no obstacle would be between A and B) by using only the known measures of AC, BC, and the angle ACB between these two sides. In other words, this method made use of one of the criteria for the congruence of triangles (SAS), a mathematical topic that the students had not encountered before. To help the students understand how and why this method would work, as well as to appreciate its interest from a mathematical viewpoint, the teacher planned a number of complementary activities for the next lesson.

The class started by sharing and compiling on newsprint an earlier homework assignment in which the students had been asked to list "at least 10 things they knew about triangles." Though few of the properties of triangles generated were relevant to solving the problem under consideration, this preliminary activity helped the students to consolidate what they knew about triangles as well as to review some relevant technical vocabulary. Furthermore, it suggested a possible connection between the lake problem discussed in the reading and properties of triangles.

After asking the students to take out part 2 of the essay and locate the section that discussed the method she wanted them to consider closely, the teacher invited a volunteer to come to the board and "act out," on the figure of the lake she had drawn there, the procedure suggested by the author. Randy was eager to volunteer and had no problem executing the various steps of the procedure described in the reading, although the teacher occasionally asked him for some clarification and justification. This demonstration seemed crucial to helping several students in the class understand how the method really worked—something a few of them admitted not having been able to understand at all from their prior independent reading of the text.

To summarize the discussion and highlight its mathematical significance, the teacher asked Randy to articulate the key steps of the procedure once again, and she then wrote on the board, "Angle, measure first side, measure second side," underneath her drawing. The property that a triangle can be reproduced when given only two sides and the angle in between was then added to the list of things the class knew about triangles created at the beginning of the class.

To further help the students understand *why* the triangulation method described in the reading worked and to make explicit its connection with the more general criteria for the congruence of triangles, the teacher also organized some hands-on activities for her students. The first of these consisted of using straw models of several geometric figures to test out the statement (made earlier in the reading) that triangles, unlike other figures like squares or other quadrilaterals, are rigid. The second activity required the students to use manipulatives in small groups to explore how a triangle could be reproduced without knowing the measurement of all its sides and angles. Though the relationship of these activities to the reading experience might not be immediately apparent, both were generated by questions raised the first time the class read the essay and enabled students both to understand the essay and to see its connection to the curriculum content of the unit.

This vignette illustrates how reading and a variety of related mathematical activities can be orchestrated to help students understand some technical mathematical content (such as distance, the congruence of triangles, or the specific properties of geometric figures) as well as their significance and potential applications. A generative reading of the text—whether it involves "saying something," "acting out" what is described in the text, or elaborating on a number of student-generated texts by compiling them on newsprint—seems essential if students are to capitalize on even intrinsically rich texts such as the essay used here. The contrast between what the class achieved when engaged in such a reading of the text and what individual students got when reading the same parts of the text independently at home suggests that reading mathematics transactionally can truly support meaningful and generative thinking.

CONCLUDING COMMENTS

A transactional perspective on reading treats reading as a far more active, generative, and social process than traditional approaches—as the vignettes reported in this article have illustrated. These examples have shown how reading (even in mathematics classrooms, and dealing with technical content!) is not always a simple matter of decoding the text and learning the content, but may instead involve reading aloud to share one's ideas, reading with a partner to make sense of a text, "acting out" what one is reading, and reading as a class to make a connection to the inquiry in progress.

Our vignettes have shown that *why* a text is read is at least as important, if not more so, than the other aspects of reading. Among the purposes for reading we have illustrated are helping students make sense of mathematical concepts or procedures, seeing connections between mathematics and real life, developing broader views of mathematics, developing strategies for sharing information, and valuing students' own ideas and those of others.

REFERENCES

Borasi, Raffaella. "The Invisible Hand Operating in Mathematics Instruction: Students' Conceptions and Expectations." In *Teaching and Learning Mathematics in the 1990s,* 1990 Yearbook of the National Council of Teachers of Mathematics, edited by Thomas J. Cooney, pp. 174–82. Reston, Va.: National Council of Teachers of Mathematics, 1990.

————. *Learning Mathematics through Inquiry.* Portsmouth, N.H.: Heinemann Educational Books, 1992.

Borasi, Raffaella, and Marjorie Siegel. "Reading, Writing, and Mathematics: Rethinking the 'Basics' and Their Relationship." In *Selected Lectures from the Seventh International Congress on Mathematical Education,* edited by David Robitaille, David Wheeler, and Carolyn Kieran, pp. 35–48. Sainte-Foy, Quebec: Les Presses de l'Université Laval, 1994.

Campbell, Douglas. *The Whole Craft of Numbers.* Boston: Prindle, Weber, & Schmidt, 1976.

Davis, Philip, and Reuben Hersh. *The Mathematical Experience.* Boston: Houghton Mifflin Co., 1981.

Harste, Jerome C., and Kathy Short, with Carolyn Burke. *Creating Classrooms for Authors.* Portsmouth, N.H.: Heinemann Educational Books, 1988.

National Council of Teachers of Mathematics. *Curriculum and Evaluation Standards for School Mathematics.* Reston, Va.: National Council of Teachers of Mathematics, 1989.

————. *Professional Standards for Teaching Mathematics.* Reston, Va.: National Council of Teachers of Mathematics, 1991.

Rosenblatt, Louise. *The Reader, the Text, the Poem.* Carbondale, Ill.: Southern Illinois University Press, 1978.

Siegel, Marjorie. "Reading as Signification." Ed.D. diss., Indiana University, 1984. Abstract in *Dissertation Abstracts International* 45 (1984): 2824A.

11

Communicating Mathematics through Literature

Ronald Narode

As AN expression of the many forms of human interaction, literature often reveals quantitative relationships that afford wonderful opportunities to communicate mathematical concepts. This article discusses some of the many instructional uses of one particularly outstanding piece of literature: Leo Tolstoy's short story entitled "How Much Land Does a Man Need?" (Tolstoy 1958). The story was first published in 1886 in czarist Russia, and it reflects Tolstoy's concern for the moral development of the Russian peasantry—emancipated in 1861—as well as promotes a national concern for their material well-being. As a noble landowner, Count Tolstoy understood the importance of mathematics to commerce. His story relates details of land prices, land areas, and boundaries, and it affords ample material for a discussion of the mathematical concepts of estimation, rates and ratios, fractions, distance and area, currency, distance and area conversions, and geometry and trigonometry. Specific suggestions about how the story may be used to teach curriculum—mathematics and otherwise—are offered here.

In keeping with Whitin and Wilde's (1992) advice that stories are meant to be enjoyed first and foremost as good literature, this story should be read without interruption for mathematical questions. The story is written in nine parts, each approximately two pages. Depending on the age of the students and the time available, some teachers may choose to read the story aloud to middle school and high school students or ask them to read the story themselves.

THE STORY

"How Much Land Does a Man Need?"

With apologies to Tolstoy, I submit a brief synopsis of the story. (Warning: Do not substitute this synopsis for the original during instruction. It omits far too much mathematical detail and art.)

76

The protagonist of the story, Poham, is a peasant who is overheard by the devil, boasting, "If I had plenty of land, I shouldn't fear the devil himself." And so the contest begins as the devil conspires to lure Poham to his demise.

First, Poham voices his dissatisfaction with having to pay fines and rents to landowners. Then the opportunity to purchase land creates discord among the peasants of the agricultural commune so that they cannot agree to buy land collectively. In competition with his neighbors, Poham manages to find the money to purchase forty acres.

Soon the problems of land ownership are reversed for Poham, and he finds himself imposing fines on his neighbors for accidental incursions onto his property. They in turn vandalize his property, and his position in the commune steadily worsens. Since many peasants are relocating to new areas, Poham anticipates staying and buying their land, until one day a passing stranger tells of good cheap land east of the Volga River. Poham cannot resist, so he sells all his holdings and moves east with his family.

There, he is given 125 acres of rich land and is told he may purchase more land at a ruble an acre. Although he considers himself ten times better off, he is tempted yet again by a passing dealer who tells Poham of a people called the Bashkirs, far to the east, from whom the dealer has purchased 13 000 acres of land for a thousand rubles: "There is more land there than you could cover if you walked a year, and it all belongs to the Bashkirs. They are as simple as sheep, and land can be got almost for nothing."

So once again Poham sets off to acquire more land. After journeying more than 300 miles (the original Russian unit was the *verst*, which is equal to 0.663 miles), he finds and befriends the Bashkirs, who offer him whatever he wishes. Of course, Poham asks for land, which the chief of the Bashkirs offers to sell for 1000 rubles a day. "A day?" asked Poham. "What measure is that? How many acres would that be?"

"We do not know how to reckon it out," says the chief. "We sell it by the day. As much as you can go round on your feet in a day is yours, and the price is 1000 rubles a day.... It will all be yours! But there is one condition: If you don't return on the same day to the spot whence you started, your money is lost." Poham agrees, calculating that he can walk thirty-five miles in a day.

Starting at dawn from the top of a hill, he sets out in an easterly direction intending to walk as large a square as he can. He walks a thousand yards and digs a hole to mark his land. He quickens his pace and after about three miles removes his undercoat and continues on. After breakfast, he walks another three miles, stops to dig a large hole, and turning sharply to the left sets out on a long, hot walk that takes him into yet better and more tempting land. Realizing that he made the sides too long and that he is far away from his starting point, Poham turns left and walks only two miles when he notices the sun halfway to the horizon. With urgency,

Poham turns directly to the hill and eventually starts running toward his destination, which he estimates to be ten miles away. His sense of panic and desperation rises as he sees the sun set over the hill before he can reach it. Just as he accepts defeat at the foot of the hill, he notices the Bashkirs cheering him on, and he realizes that the sun is still visible to those on the hill. With all his strength and resolve he runs up the hill, but his legs give way beneath him. He falls forward onto the starting point.

"Ah, that's a fine fellow!" exclaims the chief. "He has gained much land."

Poham's servant runs up and tries to help him stand up but sees that blood is flowing from Poham's mouth. Poham is dead! His servant picks up the spade and digs a grave long enough for Poham to lie in, and buries him in it. Six feet from his head to his heels is all he needs.

SUGGESTIONS TO FACILITATE MATHEMATICAL DISCUSSION

As with all good stories, the students should be encouraged to discuss their impressions and to ask one another questions about the story and the characters. This will also begin the communication process. The following are some suggestions for activities that will evoke discussion and development of students' understanding of mathematical concepts. Of course, this list is only partial and may be expanded considerably. Students could work in pairs, since collaboration requires on-task activity and communication (Whimbey and Lochhead 1981).

Prepare a timeline. Ask each pair of students to prepare a timeline that chronicles Poham's life during the period of the story. Ask students to be sure to write all the details of Poham's land acquisition, including all information about price and quantity.

Calculate price for an acre. Students should then attempt to calculate or estimate the price for each acre of Poham's land purchases, including his final purchase. This is not a trivial problem and may require students to collect information from several places in the story.

Find perimeter from area. The dealer told Poham that he acquired 13 000 acres by walking a circuit in a day. If he walked the perimeter of a square field, how far would he have walked? Is this distance reasonable?

Find average speed, time, and distance. The solution to the problem above reveals that the dealer walked approximately eighteen miles in a day. Whose distance is the more reasonable expectation of a day's walk, the dealer's eighteen miles or Poham's anticipated thirty-five miles?

Some factors to consider, besides the relative fitness and greed of the men doing the walking, are the length of the day and an estimate of average walking speed with a consideration of the terrain. Discussion and instruction should reveal that the length of the day depends on the season and on location relative to the equator. A glance at a map of Russia will locate Bashkir at around fifty-five degrees north latitude (about the same latitude as central Canada or southern Alaska). The story relates a long hot day for Poham's walk, a day most certainly in summer for such a northern clime. There is also a reference to the Bashkirs' behavior in summer. The students may estimate the number of hours from sunrise to sunset and then check an almanac in the library for confirmation.

Although estimates may be made for the average speed of walking on the flat prairie lands of Bashkir, the story describes some lands with high grasses, deep depressions, and of course the hill on which Poham started and finished his walk. Poham also stopped to dig several marker holes, and he stopped to rest and to eat.

Write a paragraph. The students should reread the section of the story that relates Poham's fatal walk. Each pair of students should prepare a paragraph relating all the details of the walk, with particular emphasis on distances, speeds, and times.

Draw a picture. The students should attempt to make a detailed sketch of Poham's route from the information they can glean from the story. The various segments should indicate Poham's activities such as digging marker holes, eating meals, removing clothing, resting, changing direction, and so on. The students should note the terrain for the various segments. Unknown distances for certain segments may be estimated or inferred.

Make several graphs. The students can now complete a detailed timeline of Poham's walk, complete with estimates of walking speeds during specified time intervals. From this timeline, the students may complete a speed-time graph and a distance-time graph. They should be asked to compare their graphs to the account related in their paragraph and to their picture. Do the different representations tell the same story? Further comparison between the two graphs could lead to a discussion on how speed, distance, and time are interrelated.

Calculate area and estimate error. The students are now in a position to calculate the number of acres Poham acquired on his walk. They may compute areas directly from their pictures and then calculate the percentage of error from the estimated distances of unknown segments. Students with backgrounds in calculus may be asked whether, and how, it is possible to compute the area from the graphs alone.

Find linear and quadratic relationships. How many acres would Poham have acquired if he had completed his thirty-five-mile square? What would have been the price for each acre in this instance? Compare these numbers to those you calculated for the dealer. Recall that Poham walked only twice as far as the dealer. How do you explain the difference in price?

Use comparative geometry. Could Poham have acquired more land if he had walked in some shape other than a square field? What would have been the price for each acre in this instance?

Relate to business. Poham planned to place 150 acres of his new land under cultivation while pasturing cattle on the rest. What percent of his land would go to farming and what percent to ranching? Do you think Poham could have managed his land given his expected resources?

Explore geography. What are the astronomical and mathematical relationships among latitude, day of the year, and length of day?

Consider earth science. While walking his circuit, Poham became alarmed when he noticed that the sun was halfway to the horizon. Was his alarm justified? How much time remained until sunset?

Use trigonometry. On his final dash, when Poham reached the hill, he saw the sun set behind it and thought that he had lost everything. But just then, he realized that the Bashkirs at the top of the hill could still see the sun. Explain how this is possible, and estimate the amount of time Poham actually had until the sunset occurred to viewers at the top of the hill. Did Poham really need to run those last fateful yards?

CONCLUDING COMMENTS

Many of the suggestions above would require extended class time and possibly several days for each. Completing *all* the suggested activities could take students weeks. Teachers may consider longer class periods using the block-scheduling approach, or they may wish to collaborate with teachers from other disciplines such as social studies, literature, science, and geography. As an extended project, students could assemble portfolios including their work on these problems, additional essays and reports related to these topics, and reflective writing on the problem-solving activities done in pairs. Students should also be encouraged to ask themselves mathematical questions of the story. This should happen early and often, since what appears here is only a sampling of questions from one excellent story.

BIBLIOGRAPHY

Raphael, Elaine, and Don Bolognese. *Sam Baker, Gone West.* New York: Viking Press, 1977.

Tolstoy, Leo. "How Much Land Does a Man Need?" In *A Treasury of Great Russian Short Stories,* edited by Avrahm Yarmolinsky, pp. 547–61. New York: Macmillan Co., 1958.

Whimbey, Arthur, and John Lochhead. *Problem Solving and Comprehension.* Hillsdale, N.J.: Lawrence Erlbaum Associates, 1991.

Whitin, David J., and Sandra Wilde. *Read Any Good Math Lately?* Portsmouth, N.H.: Heinemann, 1992.

———. *It's the Story That Counts.* Portsmouth, N.H.: Heinemann, 1995.

12

Talk Your Way into Writing

DeAnn Huinker
Connie Laughlin

THINKING and talking are important steps in the process of bringing mean-ing into students' writing. This article describes a strategy called "think-talk-write" for improving the written communication of mathematics. Adults as well as young children can benefit from using this strategy.

THE ROLE OF TALK IN LEARNING MATHEMATICS

Classroom opportunities for talk enable students to connect the lan-guage they know from their own personal experiences and backgrounds with the language of the classroom and of mathematics (Gawned 1990). The individual analyzes and synthesizes mathematical ideas as he or she identifies the aspects of the situation that are considered important and those that are not. In selecting the "right" language—words that will be recognized and accepted by others—students modify existing under-standings and construct meaning for mathematical ideas. Dialogue with others allows individuals to negotiate meaning. This access to others' thoughts permits the refinement, extension, and validation of existing ideas, as well as the clarification of partly understood ideas (Labercane 1988). In essence, this dialogue enables students to talk their way to meaning. As students talk about their experiences and test their new ideas with words, they become aware of what they really know and what more they need to learn.

Talk also fosters collaboration and helps to build a learning community in the classroom. When students are given opportunities to "talk mathe-matics" frequently, they realize their thinking is valued. This sense of community helps students feel comfortable enough to take risks. "By talking to a sympathetic partner, [students] can test ideas, explore words, experiment with different methods of organization and not lose valuable thoughts" (Reid 1983, p. 4). Students engaged in talk share a common goal of learning and become teachers of one another.

THINK-TALK-WRITE

For most children, talking is natural; writing is not. The process of talking is learned as a child and is reinforced throughout life as individuals interact with others. The naturalness and ease of talk establishes a comfortable atmosphere for students in the classroom and can be used as a valuable prewriting tool (Abbott 1991; Reid 1983). As students talk about mathematical ideas in relation to their experiences, they become able to write about these ideas.

The think-talk-write strategy builds in time for thought and reflection and for the organization of ideas and the testing of those ideas before students are expected to write. When assigned a writing task, students are often expected to begin writing immediately. The talk phase of the think-talk-write strategy allows for exploratory talk—"the process of learning without the answers fully intact" (Cazden 1988, p. 133). The flow of communication progresses from students engaging in thought or reflective dialogue with themselves, to talking and sharing ideas with one another, to writing.

This strategy seems to be particularly effective when students, working in heterogeneous groups of two to six students, are asked to explain, summarize, or reflect. Start with smaller groups and then increase the size as students become more comfortable with the process. The think-talk-write strategy is described and discussed in the following two examples.

Third-Grade Example

In an introductory lesson on the concept of division, students used counters to explore division situations in small groups and then discussed their findings as a whole group. Using the think-talk-write strategy, students explained the meaning of division.

Think

Teacher: What is division? Think for thirty seconds about what division means. No talking, just thinking. I'll tell you when the time is up.

The students are engaged in thought—a reflective dialogue with themselves.

Talk

Teacher: In your groups, take turns explaining to one another what division means. Each person should talk for about thirty seconds. When one person is talking, the others are listening. I will tell you when it's the next person's turn to talk.

The students begin talking. Here is the talk from one of the groups:

Janette: Division is putting things into groups.

Paul: When you divide, you take a group of things and split them into new smaller groups.

Nicole: The groups have to be the same in division; they have to be equal.

John: Yeah, but sometimes it doesn't come out even and you've got some left over, and that's the remainder.

Write

Teacher: Now think about what everyone in your group said and then use words—and pictures if you want—to explain what division means. Go ahead and write.

Taking turns during the talk phase is especially important at first so that all students are given an equal opportunity to talk. Taking turns ensures that each student verbalizes her or his ideas and that no student dominates. The other students can ask questions for clarification such as "What do you mean by that?" but they should not state their own ideas until their turn. Students often are not accustomed to having to talk and having to listen, so it may take some time for the process to flow smoothly. Eventually the time limit can be removed as the students learn to value the ideas of others and insist that everyone in their group talks.

Compare Janette's writing, shown in figure 12.1, and Nicole's, in figure 12.2, with what these students said. Notice that neither girl said anything about a remainder, but the concept appears in their writing. It is likely that John's statement about remainders in division helped the girls to explain more fully in their own writing the meaning of division. When students listen to one another, they use other students' talk to help clarify, add to, and extend their own thinking and reasoning.

Seventh-Grade Example

The think-talk-write strategy was used with a seventh-grade class toward the end of a measurement unit based on the *Mouse and Elephant* book from the Middle Grades Mathematics Project (Shroyer and Fitzgerald 1986). In the course of these lessons, students explored changes in area

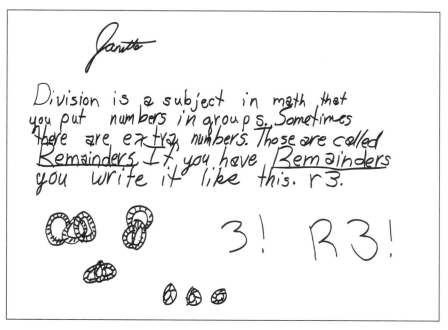

Fig. 12.1. Janette's written work (grade 3)

Fig. 12.2. Nicole's written work (grade 3)

while keeping the perimeter constant and vice versa. Students were asked to summarize their study of the relationships between area and perimeter.

Think

Teacher: For the next thirty seconds, think about the relationships between area and perimeter that you have observed over the past few weeks. No talking yet. I'll let you know when time is up. Ready? Go ahead and start thinking.

The students quietly reflect on the activities and discussions they have experienced.

Talk

Teacher: Now each person in your group is going to take a turn describing the relationships you have observed between area and perimeter. You have sixty seconds to talk while everyone else is listening. I'll let you know when it's time for the next person to talk.

The students begin talking. Initially, it is difficult for students to talk for sixty seconds; teachers might want to begin with just thirty seconds. The talk from one of the groups follows.

Jennifer: The relationship between the area of a shape and the perimeter of a shape is like the area of a shape is the inside and the perimeter is the outside.

Keisha: What I have observed is if you make one shape, you can find the area and perimeter of that same shape. You can add more tiles, and the perimeter might end up staying the same and the area changes or the area might stay the same and perimeter might change. Like for instance, we made that **U** shape in class. Many people changed the perimeter but the area stayed the same, and many people changed the area but the perimeter did not change.

George: The largest perimeter is a square. All the dimensions have to be equal. Mmm…When you start with some numbers, once you get to the maximum number, they switch, and you get the same number at the end.

Fred: A relationship they have is when perimeter of something stays the same but the area changes. One way is to make a chart with length, width, area. You have to multiply to get the area.

Write

Teacher: It's time to write. Describe the relationships you've observed between area and perimeter.

The writing of these students showed that they had listened closely to one another. First, take a look at George's writing in figure 12.3. George stated that the perimeter can stay the same as the area changes, which he heard from Keisha, but then he showed this relationship in a chart, a suggestion from Fred, and in pictures. George also decided that it was important to write, "The area is always the length times the width," which was mentioned by Fred.

Jennifer's writing in figure 12.4 shows that she also listened carefully to her group members. She had been the first to talk and only defined area and perimeter. Her writing, however, includes ideas mentioned by each of the other group members. She added a sentence about Keisha's observation about changing perimeter and constant area; she mentioned George's discovery about the maximum perimeter; and she repeated Fred's comment about the use of a chart and elaborated on his description of how to compute the area.

Incomplete notions or misconceptions occur in both students' talking and writing as a natural part of the learning process. The talking phase often helps to clarify some of these misconceptions. Some, however, are not evident until students write. Examine the chart in George's work in figure 12.3. Although

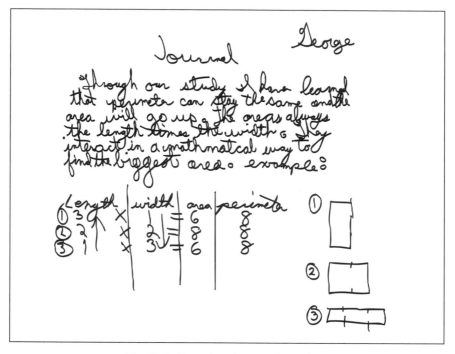

Fig. 12.3. George's written work (grade 7)

he stated that the area is always the length times the width, it is not clear how he actually computed the area, nor did he indicate whether the units are linear or square. These omissions indicate that further probing of George's understanding is necessary to reveal his perceptions of area and measurement units.

IMPACT ON LEARNING AND TEACHING

Teachers have responded positively to the think-talk-write strategy. They see improvements in students' writing and hear richer discussions among students in small groups, as the following quotations illustrate:

Now, before students write they ask, "Can we talk about it first?"

—Third-grade teacher

This strategy has really made an impact on my students. Their writing has improved, as well as their willingness to share their ideas with each other. It's also made an impact on me and how I teach. I now plan time for the students to talk to each other in small groups. I think it's so important that every student gets a chance to express his or her ideas. I used to have my students work in small groups and assumed they were talking, but they really weren't. Someone usually just took over, and some students didn't say any-

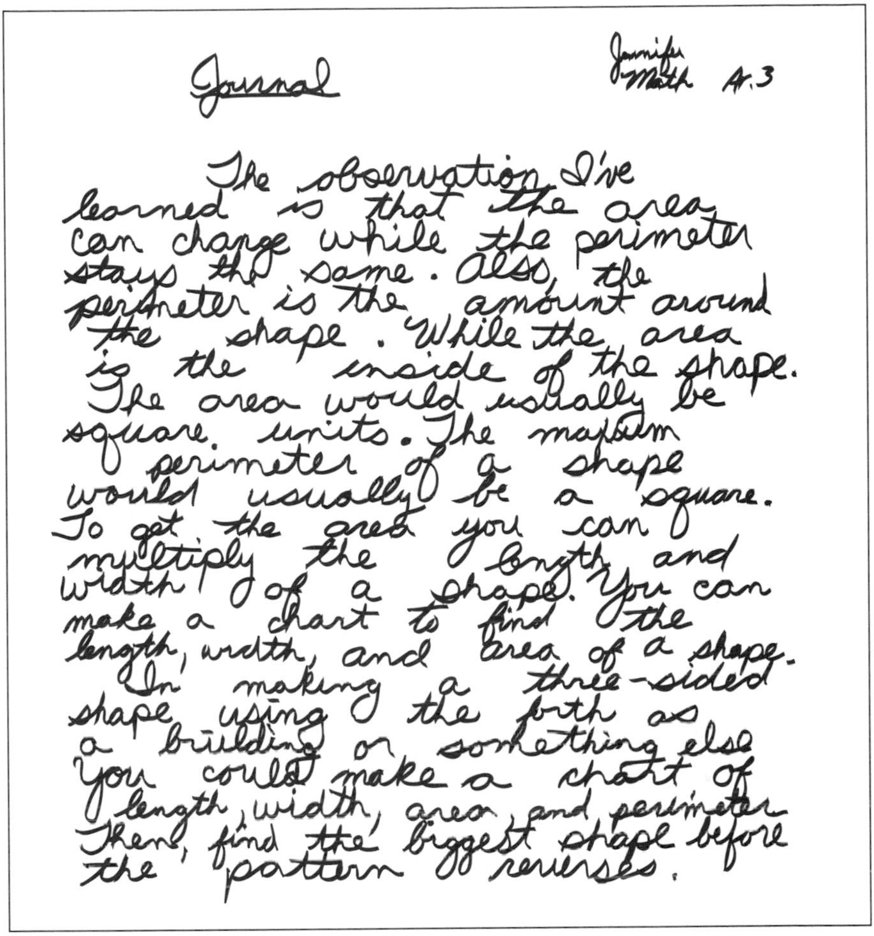

Journal

Jennifer
Math 4.3

The observation I've learned is that the area can change while the perimeter stays the same. Also, the perimeter is the amount around the shape. While the area is the inside of the shape. The area would usually be square units. The maximum perimeter of a shape would usually be a square. To get the area you can multiply the length and width of a shape. You can make a chart to find the length, width, and area of a shape. In making a three-sided shape, using the forth as a building or something else you could make a chart of length, width, area, and perimeter. Then, find the biggest shape before the pattern reverses.

Fig. 12.4. Jennifer's written work (grade 7)

thing or want to say anything. Now they all want to talk and encourage each other to talk.

—Seventh-grade teacher

Students also have expressed positive reactions to the think-talk-write strategy. Some students were asked to explain what they thought of having to think and talk before they write:

I like it when we get to talk. It helps me learn.

—Third-grade student

It helps me know it better when I can tell it to my group.

—Third-grade student

Well, sometimes I'm not sure if I really get it, but then when I hear someone else in my group explain it, it helps me understand it better.

—Seventh-grade student

CONCLUDING COMMENTS

Discussing and writing are both important and vital aspects of communication at all levels, K–12 (NCTM 1989). Classroom discussion should include cross-discussion—direct dialogue among students without the teacher's intervention (Cazden 1988). The think-talk-write strategy helps students move in this direction as they gain skill in listening to one another's comments.

In the classroom, we must ensure that talk is not neglected but given priority. The think-talk-write strategy presented here allows all students to talk out the ideas behind their thoughts before they write. Talking encourages the exploration of words and the testing of ideas. Talking promotes understanding. When students are given numerous opportunities to talk, the meaning that is constructed finds its way into students' writing, and the writing further contributes to the construction of meaning.

BIBLIOGRAPHY

Abbott, Susan. "Talking It Out: A Prewriting Tool." *English Journal* 78 (April 1991): 49–50.

Atkinson, Sue, ed. *Mathematics with Reason.* Portsmouth, N.H.: Heinemann Educational Books, 1992.

Cazden, Courtney B. *Classroom Discourse: The Language of Teaching and Learning.* Portsmouth, N.H.: Heinemann Educational Books, 1988.

Gawned, Sue. "The Emerging Model of the Language of Mathematics." In *Language in Mathematics,* edited by Jennie Bickmore-Brand, pp. 27–42. Portsmouth, N.H.: Heinemann Educational Books, 1990.

Labercane, George. *Talking the Loneliness out of Writing.* Paper presented at the annual meeting of the National Council of Teachers of English, Saint Louis, Mo., November 1988. (ERIC Document Reproduction Service no. ED 305 646)

Mumme, Judith, and Nancy Shepherd. "Implementing the *Standards:* Communication in Mathematics." *Arithmetic Teacher* 38 (September 1990): 18–22.

National Council of Teachers of Mathematics. *Curriculum and Evaluation Standards for School Mathematics.* Reston, Va.: National Council of Teachers of Mathematics, 1989.

Reid, Louann. *Talking: The Neglected Part of the Writing Process.* Paper presented at the annual meeting of the National Council of Teachers of English, Seattle, Wash., April 1983. (ERIC Document Reproduction Service no. ED 229 762)

Shroyer, Janet, and William Fitzgerald. *Mouse and Elephant: Measuring Growth.* Middle Grades Mathematics Project. Menlo Park, Calif.: Addison-Wesley Publishing Co., 1986.

13

Try a Little of the Write Stuff

Peggy A. House

I WON'T divulge how many years have elapsed since that day, but I remember the assignment as clearly as if I had completed it last week: *Write the autobiography of a nickel.* It was a turning point in my education. Even allowing for the scrawl of my eighth-grade penmanship, I produced a paper many times the length that the teacher had assigned; it seemed that once I began, there was always a new idea to pursue, a new twist to add as I played with ideas of multiples and factors, decimals and fractions, shape and size, and more—and I had never before felt so satisfied with my own accomplishment. It was then that I made two important discoveries: (1) that it was actually possible to have fun with mathematical ideas and patterns, and (2) that maybe I'd want to be a journalist when I grew up. For the next eight years I devoted considerable energy to both mathematics and journalism; in the end I gave up on journalism.

Unfortunately, far too few students have had the opportunity to encounter mathematics in a playful environment in which the sky's the limit on imagination and the product is uniquely and proudly one's own. But it is really quite easy to create such opportunities for students of any age through creative-writing activities and projects. The student writing included in this paper will illustrate some of the possibilities for promoting a deeper understanding of mathematics through creative writing.

GETTING STARTED

Teachers know that any deviation from established classroom routine is met with some degree of skepticism and apprehension on the part of the students. Introducing creative writing into mathematics class is no exception. Usually the biggest problem for students is figuring out how to get started. Once they get the project off the ground, the ideas come easily. Thus it is helpful to employ a variety of techniques for inspiring students to begin to write. Here are a few possibilities that have proved to be effective in actual classroom use:

Short, well-defined tasks

A brief, well-defined task is a good opener for writing activities. For example, *If you could be any number, which number would you be, and why?* The following are some high school students' responses:

> If I could be any number, the number I would most like to be is Zero. My reason for wanting to be Zero is that nothing would be expected of me.
>
> If I was Zero, transportation would never be a problem because I could just roll to my destination. Also, if I was Zero, I could easily visit my best friend, One, and we could make a perfect ten.
>
> Another advantage of being Zero is that, no matter how many other zeros are around, the crowd never increases. Also, no matter how big a number is, all I have to do to bring it down to my size is multiply with it.
>
> There are many uses of Zero such as basketball hoops and the rims of eyeglasses that help people see. In fact, if I am a really good Zero, I may even end up as a halo above some saint's head.
>
> *—Mike*

> My favorite number has always been Seven. Seven and I go back a long way, because I was born on July 7, 1970—7/7/70—and on July 7, 1977, I turned seven years old. As you can guess, seven is my lucky number.
>
> Let me tell you about my dream that the Duchess of Real Numbers offered to let me be any number I wanted for a day. Of course I chose seven, and the day I picked was July 7, 7777. I spent my day as Seven by going around to all the local schools and helping children with their mathematics. I was very honored to be able to help children enjoy and understand mathematics, and when I woke up I decided that I might like to be a math teacher.
>
> *—Jennifer*

> If I could be any number, I would be π, because π is one of the most distinguished numbers around. Not only that, but π goes on forever. And I would be of untold importance to geometry. Yes, being the number π would be great.
>
> As the number π, I would have many friends. I would be inseparable from the circles, and, since I would contain many digits who appear in a most unpredictable fashion, I would never become stale or uninteresting.
>
> *—Don*

Newspapers or newsletters

Students in the class can take the roles of members of a newspaper staff, each with an assigned beat to cover. Assignments and examples of some of the stories that student-reporters have filed include these:

- *News reporting*—covering the space shuttle's launch into vector space or the hijacking of the Cartesian plane
- *Business and economic news*—covering the effects of the ongoing strike by the prime numbers

- *Fashion news*—describing what the well-dressed polygon will be wearing next season
- *Home and garden column*—offering advice on the proper pruning of factor trees or the grafting of square roots
- *Entertainment news and reviews*—covering the season premiere of popular television shows like M★A★T★H or last night's rock concert by the popular Log-a-Rhythms
- *Sporting events*—reporting on the Hyper Bowl and the championship game between the positives and the negatives
- *World travel*—featuring such exotic destinations as the Islands of the Imaginaries

Other components of the student newspapers can include editorials, mathematical cartoons, crossword puzzles, want ads, and advice columns. Word-processing, graphics, and desktop-publishing programs make it easy to lay out and print attractive student newspapers that can be sent home to parents and distributed to other classes.

Advertising campaigns

Included in the student newspapers can be advertisements that the students write and illustrate. These can be as simple as the following:

Buy TIMES, the laundry detergent that multiplies as it cleans.

Are the decimal points and plugs in your car acting up again? Have them checked at FRACTION AUTO.

Or the advertisements can be intended for television (or the school's public address system) and can include jingles like the following that the students sing:

You deserve some math today,
So get up and get away
To your math class!

Posters that feature the mathematical advertisements can be designed and displayed around the classroom or school.

Titles-to-go

Frequently all that is needed to ignite a student's imagination is to suggest a title. For example, *The Day Zero Died, The Destruction of the Circles, A World without Triangles, When Numbers Are No More,* and *The Do-Nothing Number* are all student stories that evolved from suggestions embedded in titles.

GETTING CREATIVE

After students have recognized that writing and mathematics can go together in many different ways, they can be encouraged to focus their creativity on the mathematical topics that the class is studying at a particular time. For example:

Rhymes for a reason

Students can show their understanding of the topics they are studying in mathematics by crafting short poems or limericks such as these student products:

On prime numbers

"Prime numbers have only two factors,"
Said Steve, as he plowed on his tractor,
"That number and one—
"There you have them; you're done.
"See, the secret's not hard once you've cracked 'er."

On negative numbers

There once was a student named Mary
For whom negative numbers were scary.
It just didn't suit
When she found their square root
And they were only imaginary.

On squaring

Any number you pick can be squared.
Use it once and again, then they're paired.
Multiply them together.
Not hard whatsoever.
Go on now, and try. Don't be scared.

On averages

The median, the mean, and the mode:
These are all types of averages, I'm told.
They show different views,
In case one of them skews.
So learn them, all three, and be bold.

Parodies

A parody is defined as a literary work in which the style of an author or work is closely imitated for comic effect. Parodies of familiar works, especially nursery rhymes, fairy tales, and children's stories, have proved to be popular with students. For example:

Baa, baa, black sheep,
Can you graph a line?
Yes sir, yes sir,
I do it all the time.
Changes in y and x give slope,

And then I add the *b*
To find the *y*-intercept and thus
Complete it, don't you see?

Space does not permit the reprinting of long stories, but a sampling of work produced by students includes the following:

- *The Wizard of Oz* became *The Wizard of Odds*.
- *Alice in Wonderland* gave rise to *Alice in Numberland* and to *Alice in Cartesian Land*.
- *Snow White and the Seven Dwarfs* begot *Snow One and the Seven Prime Numbers* and *So-Bright and the Seven Triangles*.
- *Cinderella* emerged as *Squarella*, as *Circlerella*, and as *Equilaterella*.
- *The Ugly Duckling* was reincarnated as both *The Ugly Numberlings* and *The Ugly Triangle*.
- *Winnie the Pooh* became *Winnie the Two*.
- *Puss in Boots* yielded *Pi and Roots*.

Spin-offs

Spin-offs are created as extensions of other works. One way to develop spin-offs is in connection with books or films that can be read or shown in class. For example, after reading *Flatland* (Abbott 1952), students might write about what it would be like to play their favorite sports in Flatland. Movies and television shows frequently are selected by students who create spin-offs to match. The Planet of the Apes series inspired the writing of *The Planet of the Zoids* and *Beneath the Planet of Polygon*. The Star Trek television series inspired *Star Tech: The Next Iteration*. *Little House on the Plane, The Adventures of Huckleberry Function,* and *The Variable Who Came in from the Cold* are other examples of how students have cleverly related mathematical ideas to familiar shows and literary styles.

Songs

One of the most enjoyable variations of mathematical creative writing has proved to be the writing of new lyrics for familiar songs because once they are written, students can have classroom sing-alongs. Also, whereas poems and stories are probably best written alone, song lyrics lend themselves particularly well to production by pairs or small groups of students. The following are but three verses of a much longer song adapted by a group of students:

My Favorite Things

Arc sines and cosines, a heavenly vision;
Cubic equations, synthetic division;
Functions with zeroes that make my heart sing:
These are a few of my favorite things.

Diff'rentiations that run to five powers;
Homework assignments that run to five hours;

Inte-gra-tion and the madness it brings:
These are a few of my favorite things.

Inverse relations, continuous functions;
Logical thinking, the rules of disjunctions;
Infinite series with limit and bound:
These are the things that will make my heart pound.

Problem stories

Problem stories combine writing with the presentation of problems or puzzles developed by the student. Students can begin by writing short "Who Am I?" riddles and later work on longer, more involved stories. An added appeal of problem stories is that they can be written for peers or younger students to solve.

CONCLUDING COMMENTS

Students who have participated in writing activities such as these report that at first they were confused about "what this has to do with math," but once they got into the activities, they began to enjoy them more and more and to see the mathematical connections. (For more examples of student writing, see House and Desmond [1994].) On the days when writing assignments were due, the students were eager to hear classmates read their works. Bulletin board displays and booklets of the collected writing of the students have been received with enthusiasm. Many students also produced clever illustrations to enhance their stories, poems, and essays.

Writing assignments should be tied to class discussions of related mathematics. For example, the stories frequently contain clever mathematical puns that the students can identify and explain. Fantasies like Abbott's *Flatland* (1952) can spark a discussion of dimension and of what it might be like to exist in a space of other than three dimensions. Writing and then talking about a particular number or geometric shape can help students focus on special properties of that object.

But the most significant outcome of the writing activities that we have done with students has been the recognition that mathematics has a playful human dimension, a recreational fascination as well as a serious purpose, creativity as well as precision. Students have had fun with their mathematics, and they came away not only with a deeper understanding of the mathematical topics but also with the realization that maybe they have the "write stuff" to push deeper into the subject.

REFERENCES

Abbott, Edwin A. *Flatland*. New York: Dover Publications, 1952.

House, Peggy A., and Nancy S. Desmond, eds. *Mathematics Write Now!* Dedham, Mass.: Janson Publications, 1994.

14

Developing and Assessing Mathematical Understanding in Calculus through Writing

Joanna O. Masingila
Ewa Prus-Wisniowska

W HEN students are encouraged and required to communicate mathematically with other students, with the teacher, and with themselves, they have opportunities to explore, organize, and connect their thinking. We have used writing to promote communication in our classes. Writing can help students make their tacit knowledge and thoughts more explicit so that they can look at, and reflect on, their knowledge and thoughts. For teachers, writing can elicit (*a*) direct communication from all members of a class, (*b*) information about students' errors, misconceptions, thought habits, and beliefs, (*c*) various students' conceptions of the same idea, and (*d*) tangible evidence of students' achievement.

In the following sections we will discuss and provide examples of how we have used various writing methods to develop and assess mathematical knowledge. The activities suggested here offer a way to get started using writing in calculus classes; we encourage teachers to adapt these ideas for any mathematics content and level.

DEVELOPING MATHEMATICAL UNDERSTANDING

Helping students build mathematical understanding entails helping them build a mental network of representations. Writing is a powerful tool for building such a network (Hiebert and Carpenter 1992, p. 67):

A mathematical idea or procedure or fact is understood if it is part of an internal network. More specifically, the mathematics is understood if its mental

---

Ewa Prus-Wisniowska died in Szczecin, Poland, while this article was in production.

representation is part of a network of representations. The degree of understanding is determined by the number and the strength of the connections. A mathematical idea, procedure, or fact is understood thoroughly if it is linked to existing networks with stronger or more numerous connections.

Writing is a "meaning-making process that involves the learner in actively building connections between what she's learning and what is already known" (Mayher, Lester, and Pradl 1983, p. 78).

Writing from a Prompt

One way to engage students in writing is to give a prompt to which they respond. Through carefully chosen prompts, teachers can encourage students to connect new learnings with prior knowledge and to develop new understandings about concepts. Two prompts we have used are listed here:

1. Explain the similarity in, and difference between, using a vertical line to test whether a rule is a function and using a horizontal line to test whether a function is a one-to-one function.

2. For a given partition P of the interval $[a, b]$ and a positive function f on the interval $[a, b]$, you can form upper and lower sums that are particular approximations of the exact area under the curve $y = f(x)$ from a to b. These approximations, as imperfect estimates, can be improved by adding points to the partition P and obtaining new upper and lower sums—numbers that are closer to the exact area. Explain why this is so.

Students' written responses offer insight into the understandings they are developing. Dylan's response to the first prompt indicates a very concrete and procedural understanding of the line tests: "In both of the tests the idea is to see if the line intersects the rule more than one time. For example, $f(x) = |x|$ passes the vertical line test but not the horizontal line test."

Jackie was able to compare the two tests more generally and see similarities and differences:

> Using a vertical line to test whether a rule is a function and using a horizontal line to test whether a function is one-to-one is similar in that both tests are checking to see whether a selected number is paired with more than one number, as a coordinate pair. They are different because the testing whether a rule is a function means seeing if each domain value is paired with only one range value while testing whether a function is one-to-one means seeing if a range value is paired with more than one domain value.

The second prompt was given after the first class on the Riemann integral. We set up the introduction of the definite integral by discussing it as a tool for finding the area between the graph of the function $f(x)$ and the x-axis. Then the definition of the definite integral was given, and the students worked on a few examples in which they were asked to estimate the values of the definite integrals. We posed this prompt to help students reconsider and make sense of the definition of the integral.

Pam attempted to explain the procedure using a visual interpretation:

> By adding points to the partition one makes the area of the regions between the upper and lower sums smaller as illustrated below. The more points, the closer these approximations come to the actual value of the area under the curve.

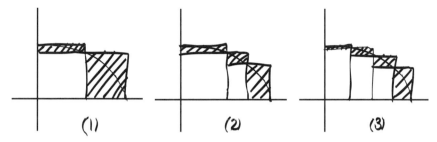

The shaded region represents the margin of error. As you can see, the graph with more points—(3)—has much less shaded area—showing it is the most accurate.

Philip revealed a deeper understanding of the integral as a limiting process:

> As the number of points in the partition increases, the number of rectangles increases too, which makes them more closely follow the curve. Therefore, the upper sum decreases and the lower sum increases, until finally, the actual area under the curve is "trapped" between these estimated areas, and the definite answer is reached.

Writing from a prompt allows students to express and teachers to see the personal nature of making sense of a new idea. A new idea makes sense for a student if he or she is able to link it with a network of mental representations. Writing from a prompt encourages students to forge new links and think reflectively about the links they have already made.

Writing in Performance Tasks

Performance tasks involve students in demonstrating and communicating their understanding. These tasks allow the teacher to examine the students' process as well as product. Whereas writing from a prompt involves students in explaining their thinking about a past mathematical activity or idea, performance tasks require that students first engage in doing a task and then communicate their understanding and thinking. Well-chosen tasks can help the teacher communicate to students the context in which mathematical understanding is constructed, and students come to understand mathematical ideas not by hearing detailed lectures about predetermined mathematical relationships but rather through active, constructive efforts.

Individual tasks

We have used the following individual performance tasks:

1. Is it true or false that if

$$\int_a^b f(x)dx = 0,$$

then the function f itself is equal to 0 on the interval $[a, b]$? If your answer is true, explain why you think so. If your answer is false, give an example of a function that is not identically equal to 0 on $[a, b]$ but such that

$$\int_a^b f(x)dx = 0.$$

2. Is

$$\int_{-1}^1 e^{x^2}dx$$

positive, negative, or 0? Justify your answer.

These tasks require that students conduct mathematical investigations before communicating their reasoning in writing. Many students found examples of odd functions and explained why they give the value 0 when they are integrated over an interval symmetrical about (0, 0). However, the responses of other students revealed their misconceptions and confusion concerning the definite integral. For example, Robert wrote the following response:

Take the function $f(x) = x^3$ when $f(x)$ is integrated from –1 to 1;

$$\int_{-1}^1 x^3 dx = \frac{x^4}{4}\bigg|_{-1}^1 = \frac{1}{4} - \frac{1}{4} = 0.$$

The answer is zero, however this function does have some area under the curve from –1 to 1 but would not be accounted for if done in this manner.

Robert's confusion indicated to us that his understanding of an integral was built on the case of a positive function and the area beneath its graph. In the process of applying the integral for finding areas and volumes, he seems to have developed a sequence of invalid implications, namely, that if the integral is 0, there is no area between the curve and the x-axis and all points of the curve are located on the x-axis.

Another student's response revealed her misunderstanding of the relationship between the integral and the integrand: "f has to be zero because

$f(x)$ is said to be equal to zero from $[a, b]$." Her answer to the second problem,

$$\text{``} \int_{-1}^{1} e^{x^2} dx$$

must be zero because

$$\int_{-1}^{1} e^{x^2} dx = e^{(-1)^2} - e^{(1)^2} = e - e = 0,\text{''}$$

confirmed that this student made a false connection between the integral and the integrand.

Group tasks

We have frequently used group performance tasks because we have found them to be especially helpful in developing students' mathematical understandings. During a group performance task students have to (*a*) reflect individually and together about the task, (*b*) communicate about the task with one another, (*c*) listen to what other group members are communicating, (*d*) communicate their thinking to other group members and often convince others of their plan or solution, (*e*) assess someone else's solution at hand, and (*f*) plan together how to communicate the group's solution.

One group performance task involves the concept of proof. Mathematical proof and reasoning are often not used in depth in beginning calculus classes. Students are much more accustomed to methods emphasizing computations and symbol manipulation than to those emphasizing conjecturing and justifying.

We wanted to emphasize that proof is a continuous process involving posing conjectures, modifying them, carefully reasoning about facts, refuting, and finally accepting a proof. We thought this activity would also furnish feedback about the areas that cause students difficulty in understanding and writing proofs. To begin the activity, we gave our students the following task:

> Assume that $y = f(x)$ is a one-to-one function and differentiable for every x for which the function is defined. Is the function $y = f^{-1}(x)$ differentiable everywhere?

We then analyzed the students' answers and selected several that we used for the group performance task. For the group task, we asked the students to examine the answers to the foregoing question and evaluate the conjectures made or the appropriateness of the examples used. Figure 14.1 shows two of the problems that we used for this group performance task.

From the group's responses to this task, we found that students were not ready to perform even a simple mathematical proof on their own. No group response was totally correct. The students did not see the significance of

Problem 1

Recently you considered the following question:

Assume that $y = f(x)$ is a one-to-one function and differentiable for every x for which the function is defined. Is the function $y = f^{-1}(x)$ differentiable everywhere?

Here is one student's response:

If you graph $f(x)$ and it is continuous and differentiable everywhere, then f^{-1} must also be continuous and differentiable, since $f^{-1}(x)$ is simply a reflection of $f(x)$ about the line $y = x$ and $f^{-1}(x)$ is simply $1/f(x)$.

Your Task

The student made four statements about the inverse of a differentiable function f:

Which of the four statements are true? Fill out the chart below.

1. f^1 must be continuous.

2. f^{-1} must be differentiable.

3. f^{-1} is a reflection of f about the line $y = x$.

4. $f^{-1}(x) = 1/f(x)$.

	True	False
1.		
2.		
3.		
4.		

Now pick one of your own conjectures from the "False" column and try to find an example that illustrates the falseness of the statement. For example, if you identified Statement 1 as a false statement, try to find an example of a one-to-one differentiable function such that its inverse is not continuous.

Problem 2

Recently you considered the following question:

Assume that $y = f(x)$ is a one-to-one function and differentiable for every x for which the function is defined. Is the function $y = f^{-1}(x)$ differentiable everywhere?

Here are two students' responses:

1. Yes, if $f(x) = 3x + 4$. It is differentiable, $f'(x) = 3$, the inverse is also differentiable because it would still be a function like $3x + 4$—only different but still differentiable.

2. Yes, $y = f^{-1}(x)$ will be differentiable. For example, the graph of $f(x)$ and its inverse are pictured here.

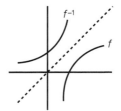

It is possible to draw a tangent line to any point on $f^{-1}(x)$.

Your Task

Are these two examples relevant to the hypotheses posed? Using only these two specific examples, can you generalize that the inverse of a differentiable one-to-one function is also differentiable?

Fig. 14.1. Two problems used for a group performance task

proof by counterexample: (*a*) they did not understand the process of refuting with a counterexample (e.g., the counterexample must satisfy all the assumptions but violate the conclusion), and (*b*) they did not appreciate the role of a counterexample (e.g., students were confused in determining whether a simple counterexample is sufficient to reject a conjecture). Many students were not aware of the distinctions between evidence and proof. For example, one group wrote that "more examples need to be shown in order for the hypothesis to be proven. In order to prove a hypothesis like this you must first try to prove it wrong. If you can't prove it wrong, then it must be true."

The task also revealed that the students did not understand the concept of inverse function. Several groups drew parabola-like functions with a horizontal tangent to support their claims that the inverse does not have to be differentiable. Thus, they ignored the fact that in order to have an inverse, a function must be a one-to-one function. The students were able to repeat the definition of an inverse function, but they associated it with the reflection of the graph about the line $y = x$ without noting whether or not this procedure generated a function.

As a result of this group task and the follow-up class discussion when the graded responses were returned, the students had the opportunity to rethink their ideas about inverse functions and mathematical proof. Being forced to communicate their understanding about these concepts brought the students' misunderstandings out into the open, where they could be examined and worked out.

Assessing Mathematical Understanding

Because we expect our students to communicate their mathematical understandings to us, we also need to find effective methods of assessing and communicating our assessment to them. Stenmark (1989, p. 4) noted that "the purpose of assessment should be to improve learning." We would add that assessment can also be used to promote and facilitate communication. If students' work is assessed in a way that communicates the areas in which the students' understandings are deep and valid mathematically and the areas in which their understandings need to be reexamined, then the assessment mechanism can reinforce and be part of written communication.

We assessed the tasks described in this paper holistically, using a generic scoring scheme (see fig. 14.2) that we adapted for each problem. In general, we have followed the guidelines discussed by Charles, Lester, and O'Daffer (1987) and Stenmark (1989) and developed scoring rubrics that were meaningful for each task.

Writing from a Prompt

For the problem about the vertical and horizontal line tests, we adapted the scoring scheme in the following way: (*a*) students who gave

---

Unsuccessful Responses

0 points: Work is meaningless; students make no progress; students fail to indicate which information is appropriate to the problem.

1 point: Students make some initial progress, but the response is incomplete because they reach an early impasse or misinterpret ideas involved in the problem.

2 points: Response is in the proper direction, but students make major errors; the response displays some substance in the sense that key ideas are identified but the relationships among them are not explained.

Successful Responses

3 points: Students work out a reasonable solution, but minor errors occur in notation or form; some explanations may lack precision, but no substantial errors occur in students' reasoning.

4 points: Solution is complete; all important ideas are identified, and their significance and relationships are discussed.

Fig. 14.2. A generic scoring scheme

no response or give only invalid examples to illustrate either the similarity or difference between the two line tests received 0 points, (*b*) students who provided a valid example to illustrate the similarity or difference and made no attempt at, or gave an invalid example for, the other or gave invalid explanations for both received 1 point, (*c*) students who gave valid examples for both the similarity and difference or furnished a valid explanation for one of them and an invalid example for, or made no attempt at, the other received 2 points, (*d*) students who provided explanations (one valid, one invalid) for both the similarity and difference or gave an explanation for either of them and a valid example for the other received 3 points, (*e*) students who furnished valid explanations for both the similarity and the difference between the two line tests received 4 points.

Through the scoring scheme we wanted to communicate to students the importance of justifying through explanations rather than simply by example. Students were, however, given some points for valid examples. Often when we had students write from prompts, we gave the students their graded response and a copy of the scoring scheme and asked them to revise their written responses and give a self-assessment of their revised response using the scoring rubric.

Writing in Performance Tasks

For the proof task, we assessed the group responses using the generic scoring scale directly. We assigned scores of 2 to two group responses; they displayed some substance, although major errors occurred. For example, one group working on Problem 1 filled out the chart correctly and then chose the most difficult conjecture to refute, namely, that the inverse of a differentiable function must be differentiable. They used the argument that the horizontal tangent for the original function yields a vertical tangent for the inverse; however, the example they used to illustrate their explanation was not invertible.

In designing this task, we overlooked the possibility that groups might not be able to work out a solution and thus they would be hesitant to write anything. In thinking about future performance tasks, we recognize the need to encourage students to describe all attempts they make regardless of whether a satisfactory solution is found. Such description is an important element in communication, and communicating the solution process in detail certainly helps the teacher and student in assessing the student's work.

As a follow-up to this task, we discussed in class several examples in which building a solution only on the observed pattern for a few specific cases leads to a failure to solve the problem. Also, we created new tasks that enabled the students to realize that a certain result is extremely plausible (e.g., moving from Rolle's theorem to the mean-value theorem). And we had the students work in small groups to discuss the relevance of all the assumptions of the proposed theorem by showing that violating them results in producing a counterexample. Discussing the group results as a whole class allowed the students to generate intuitive ideas about the proof of the theorem.

Concluding Comments

Writing in the mathematics classroom can be a means through which mathematical understanding is developed and assessed. Mathematical understanding can be developed through students' critical thinking put into words. The act of communicating with self and others can help students form bonds with existing networks of knowledge.

Only through modes of communication can teachers assess students' mathematical understandings. Likewise, only through communicating about the assessment can students understand how the teacher views their understandings. Thus, writing by teachers and students is both required and promoted by the task and assessment tool.

References

Charles, Randall, Frank Lester, and Phares O'Daffer. *How to Evaluate Progress in Problem Solving*. Reston, Va.: National Council of Teachers of Mathematics, 1987.

Hiebert, James, and Thomas P. Carpenter. "Learning and Teaching with Understanding." In *Handbook of Research on Mathematics Teaching and Learning,* edited by Douglas A. Grouws, pp. 65–97. New York: Macmillan Publishing Co., 1992.

Mayher, John S., Nancy B. Lester, and Gordon M. Pradl. *Learning to Write: Writing to Learn.* Upper Montclair, N.J.: Boynton/Cook Publishers, 1983.

Stenmark, Jean Kerr. *Assessment Alternatives in Mathematics: An Overview of Assessment Techniques That Promote Learning.* Berkeley, Calif.: Regents, University of California, 1989.

15

Is Anybody Listening?

Susan E. B. Pirie

For more than ten years now, the role of speaking has been emphasized in mathematics learning and teaching, but the role of listening has received scant attention. This article focuses on the function and nature of active listening, particularly as it applies to the teacher. Through an analysis of transcripts of teachers and students in the classroom, the article highlights what such listening is, and when and how it can occur. It aims to draw attention to some instinctive teacher activities that are seen as good practice and to question whether these activities are *always* appropriate. Above all, however, it aims to make the reader aware of the power of teacher listening (this is not to deny the need for students to listen, too, of course) to enhance student learning.

In all this, it is possible to lose sight of the fact that communication is a two-way process. Communication needs a listener as well as a speaker. Mathematical communication can take place effectively only if *all* participants are prepared to adopt *both* roles, to listen *actively* as well as to talk. Teachers who obviously engage in active listening demonstrate the value that they place on the students' contributions. This attitude in its turn can create an environment in which students are prepared to give consideration to *one another's* utterances rather than treat the teacher as the only source of correct answers. In a traditional classroom, the roles of teacher as knowledge giver and student as recipient of this wisdom tend to lead to the students' undervaluing of their peers' ideas.

LISTENING TO "WHAT"

One of a teacher's aims is likely to be to help students use mathematical language appropriately. A powerful way to do this is to use that language in ways that help the mathematical words become part of the common language of the classroom.

In Moira's seventh-grade classroom, students worked in pairs on a set of materials designed to lead them to an understanding of the solution of linear equations. The task was initially presented in the form of pictures, such as the one in figure 15.1, with the students being asked to "guess the weight of the tin." The aim, however, is for the students to come up with some rules, so that they can give a friend in the next class a strategy for making a winning "guess" every time. Damian and Brent thought they had "cracked it" and called the teacher over.

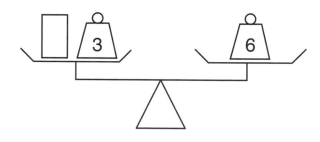

Fig. 15.1

Teacher: So what do you think the rules are?

Damian: Well, you get all the tins on one side and all the weights on the other.

Teacher: But suppose I don't know how to do that?

Damian: So, you've got to take the same off both sides. The same number of tins, or weights off this side and off there so you can get rid of them.

Teacher: OK, you do the same thing to both sides to get all the tins on one side and the weights on the other, and then what, Brent?

Damian: You just share.

Brent: You share it to get 1 tin, like if it was 5 tins and 10 kilograms then it would be 2 kilograms, 10 shared into 5.

Teacher: Fine, so you get all the tins on one side and then you can divide to get the weight of 1 tin. Well done.

It certainly *sounds* as if the students have cracked it!

A clear example of an attempt to help a student become familiar with mathematical language without making a big issue of it occurs when the teacher substitutes *divide* for *share,* using mathematical terminology instead of repeating the natural language of the student. More normal expressions than *share it* and *share into five* might be *shared* by *five* or *shared* between *five* that make it clearer which number is divided by which, but the student's intent here seems to be clear. It is in fact Brent's language, "ten shared into five," however, that reveals his underlying misconception about division. Brent seems to be trying to verbally combine the two images for division—that of *sharing* ("10 *shared* by 5") and that of *grouping* ("10 *grouped* into 5s")—and he was later heard dealing with "9 ÷ 18" by saying "nine shared into eighteen, into eighteen, ... is two," and then for "24 ÷ 6," "twenty-four shared into six is four," indicating the lack of a good understanding of either of the ways of approaching division. (Throughout this paper, the word *image* will be used to mean an idea, mental or pictorial, for a meaning for the concept. See Pirie and Kieren [1994].) His language is such that he is always able to justify mentally, through his verbalism for the division symbol, his preferred strategy of dividing the smaller number into the larger—it depends on whether he stresses the idea of *going into* or *sharing between*. Simply providing him with the mathematical term *divide* did not resolve this misunderstanding. It is the all-important prepositions *by* and *into* that make the difference: "ten divided *by* five" for "$10 \div 5$" and "ten divided *into* five" for "$5 \div 10$." The problem is, of course, that in ordinary language, the student might well say, "ten divided into five" as in "a cake divided into five"—the reverse meaning.

Returning to the substance of the actual task, if Brent and Damian have in fact understood the rules for winning at equation solving, how does one explain their thinking when, later, working on written equations, Damian looks at a series of questions of the form $5t - 3 = 3t + 5$ and says, "Two t equals two, so a tin is 1 kilogram"? The answer lies in the teacher's paraphrase of Damian's statement of the rule. Damian's actual words were, "take the same off," and his reasoning—that he can and does articulate—follows: "Take off 3 is 5 "t" take away nothing equals 3 "t" take off 3. That's 2. Take off 3 tins [implying from each side] is 2 tins equals 2."

Several months later, when the class returned to the topic of linear equations, Damian was still talking in terms of "taking off the same things." His language had originally developed and been reinforced by an image of the physical situation of weighing, in which all one can do is "take things off." Above, we see him saying "take off three," since he cannot physically or pictorially represent *negative three*. When the teacher paraphrased this as "do the same thing," neither he nor Damian was really listening to the words of the other, each was making sense on the basis of his own existing understanding, yet it was the precise words that were the key to Damian's problem: he could not model negative weights with his image for linear equations.

LISTENING TO "WHEN"

Listening involves more than just *hearing*—however precisely—the words that are said. Frequently, the listener needs to pay attention to context as well as words, as the next protocol illustrates. In a different classroom, two students, Kevin and Dave, were also working together on linear equations, this time making tables of values and plotting graphs. One of the equations given on the worksheet was $y = 2x + 3$.

Kevin: You fill it (*referring to their table*).
        x—one, x squared—two, x squared plus three—five.
        x—two, x squared—four, x squared plus three—seven.

x—three, x squared—six, x squared plus three—nine.
x—four.... Wait, we need some minuses ... er.... Do minus three.

Dave: Easy-peasy. Minus three, x squared—minus six—minus three minus two, x squared—minus two, ... no ... minus four—minus one minus one, x squared—minus two—plus one.

At this point the teacher passed by and hearing "x squared," stopped.

Teacher: Hang on. Go back a bit. You said, "If x is three, then x squared is six?"

Kevin: Yeah.

Teacher: OK. Hang on a minute.

He then proceeded to draw a three-by-three square, divide it up into unit squares, and count them.

Teacher: So it's nine, isn't it.

Kevin: What?

The teacher repeated the explanation and demonstration with a four-by-four square and a five-by-five square.

Kevin: Oh, yeah. (*Doubtfully*)

Teacher: Can you carry on now?

Kevin: Yeah.

The teacher moved away.

Kevin: What's he on about? I know all that stuff about areas and counting things.

Dave: Yeah, what we got here's lines.

Kevin: Just ignore him. Your turn. You do x squared minus three (*writing $2x - 3$*).

The teacher here has heard a misuse of language—*squared* for *twice*—but interpreted the problem as one of misunderstanding. Without pausing to examine the context in which the students were using the word *squared*, he stepped in to explain a concept that he assumed was not well understood. Perhaps fortunately, the intervention was seen as irrelevant by the students and did not interfere with their continued mathematical work. They possessed a shared understanding of the mathematics that they were doing, and they used language that *seemed to them* appropriate to the task. All that was needed from the teacher was the remark that "we read '$2x$' as 'two x' and 'x^2' as 'x squared.'" The problem was one of language and not of understanding, which would have been revealed if the teacher had listened to the way in which the students' language was being used. (All the episodes presented in this paper have been worked on by the teachers themselves. Much of what they noticed was not noticed when the talk was heard for the first time. There is, therefore, no implied criticism of the teaching reported here. On the contrary, the intention is to offer some of the insights that these teachers have gained from listening to their own teaching.)

The next episode illustrates the problem of differing contexts: that of the teacher was mathematical, that of the student was the real world. The teacher was exploring the students' understanding of the density of numbers. The class had been working on decimals recently, so he chose this topic as a way to approach the ideas.

Teacher: (*Writing "3.45" and "3.46" on the chalkboard*) Natalie, how many numbers are there between 3.45 and 3.46?

Natalie: (*Silence and a puzzled expression*)

Teacher: How many numbers could there be between 3.45 and 3.46?

Natalie: (*Studies the chalkboard hard for a long moment*) About nine or ten, I think, but there could be a lot more. You might be able to put a lot more.

Later in the day the researcher replayed the video snip to Natalie and asked her what she had meant by her response.

Natalie: Well, it was hard to tell from the back.

Researcher: Well, OK, how about if I write them down for you. (*Thinking that perhaps the student had not been able to see the chalkboard clearly, the researcher writes "3.45   3.46"*)

Natalie: (*Starting to write "1, 2, 3, 4" in the space between the two decimals*) But you see if he wrote small he might have been able to get lots more in. See I didn't understand when he said, "How many are there?" 'cause there weren't any!

At least this student was really listening to the teacher's words!

Teachers must, of course, constantly make assumptions about the meaning of what they hear and guess at what the student might be trying to express in the given situation. Struggling students can be given confidence by the teacher's supplying appropriate terms for concepts that the students are trying to communicate. The phrase "I think I know what you mean" followed by a summary or a rephrasing of the student's words is one way of supporting the student who is coming to grips with a new idea. The inclusion of the phrase "I think" is all important because it conveys that the teacher is listening and trying to make sense of what the student is thinking, not merely checking the student's version against the teacher's own understanding. It also opens the way for the students to try to articulate their thinking through mathematical language *and* to question the teacher's comprehension of that thinking. Without this openness to the student's explanation, however, there lies a danger in paraphrasing—namely, that the teacher's version may not be what the student really meant. Consider the following example:

Teacher: Who wants to do $10 - (5 - 3)$?

Sally: Subtract the 3 and the 5 and get 2.

Teacher: Yes, subtract the 3 from the 5 and get 2, so your answer is 8. Well done!

> *Sally:*   *(With a look of total incomprehension on her face, sotto voce to herself)* What? Where did the 8 come from?

The correct answer of "2" led the teacher to the belief that Sally knew how to work with the brackets in the expression, and the teacher was anxious to praise her accordingly. The teacher's paraphrase left Sally totally bewildered, however, since Sally had subtracted the 3 and the 5 from the 10 and 2 was thus her answer to the *whole* question, not just the part in brackets. Again, a subtle, one-word change carries the crucial distinction: "subtract the 3 *and* the 5" becomes "subtract the 3 *from* the 5."

Another response that will clarify the intended meaning behind a student's words is for the teacher to feign a lack of understanding with a phrase such as "I am not sure that I have quite understood." It can be quite revealing to pretend to be slower at grasping what the students are trying to communicate, and thereby push them toward a way of expressing their ideas with greater clarity.

LISTENING TO "HOW"

As has already been seen, listening to the attempts of students to verbalize their mathematics often allows teachers to predict or recognize students' misunderstandings. As a diagnostic technique, getting students to "talk math" can be a powerful tool. Greg seemed to be having some trouble with decimals. Much of the time his work was correct; he was confident in his own mathematics, but occasionally strange and seemingly random errors appeared. The teacher decided to set aside some time to talk with Greg when the rest of the class was otherwise occupied. She started by trying to get a feel for Greg's understanding of what a decimal is.

> *Teacher:* Now Greg, you seem to be on top of decimals, but I'd like to explore some decimal work with you just to see how well you are doing. OK? Can you write these down for me: Two point six, nought point three two, one point nought four.

Greg wrote all three swiftly and correctly.

> *Teacher:* Right, could you just tell me how you would say that number there *(indicating "0.45")*.
> *Greg:*   Um ... nought point four tenths and five hundredths.
> *Teacher:* Nought point four tenths and five hundredths, right?

Note that the teacher does not *paraphrase Greg's words.*

> *Greg:*   Yeah.
> *Teacher:* How about this number here *(indicating "0.6")*?
> *Greg:*   Nought point six hundredths, no, nought point six tenths.
> *Teacher:* Mm, which is bigger?
> *Greg:*   The bottom one.

Teacher: Right, and how can you always tell?

Greg: Because ... um ... it's according to the first number in line with the decimal point, after it.

Teacher: Ah, right, I see what you mean, very good.

The teacher continued by asking Greg whether he thought that it was possible for there to be a number or numbers between 0.65 and 0.66. Greg first offered "nought point six five five" but went on to say that there were "as many as you like—you can always put in another half, a sort of point five like 0.655555 is between 0.65555 and 0.65556." Apart from an odd reading of decimals at the start, presumably because he was not sure what the teacher wanted him to say, Greg appeared to have a very good grasp of the concept. The teacher left him, mystified as to the cause of his occasional errors.

A few weeks later, the class was being tested on scale reading orally and individually so that reading ability would not be a factor in the assessment. After reading the scales from a thermometer, a dial-scale pan, and a meter rule, the teacher produced a scale running from 0 to 3 drawn on a strip of card stock and divided into tenths.

Teacher: Let's have a look at this one. Could you just read that scale for me, please (*indicating "1.6"*)?

Greg: Um ... er, one inch ... I think.

Teacher: Well, don't worry about whether they're inches, just read it.

Greg: OK. One inch and one, two, three, four, ... and one seven, one and a seven ... um.

Teacher: A seventh?

Greg: No, a seven. One and a seven.

Teacher: Well, can you read it as a decimal number, as something ...?

Greg: (*Confidently and without hesitation*) One point seven (*the teacher leaves a pause*), ... nought point ... yes, one point seven.

Teacher: OK. Can you show me how you got that one point seven?

Greg: Well, that one was a whole number, so it would go in the units which would be before the decimal place.

Teacher: Yes, right.

Greg: And that one was less than a whole one so it would go in the tenths, which was seven so it would be one point seven.

Teacher: Right, how did you get the seven?

Greg: By the ... um, by the marks here, you got one, two, three, four, five, ... oh, six. I miscounted that.

Teacher: (*Relieved that Greg was not counting the initial mark for "1" as the first of his "tenths" and smiling*) Oh, that's all right, one point six. We'll call that one one point six, then. How about this scale now (*indicating a similar card, divided into fifths*)? I'll give you a question on this scale now, all right? It's a different scale. And if I mark off an arrow there (*indicating "1.6"*), what would you say that was?

Greg: One point ... one point three.

Teacher: One point three, OK. And how about this one (*indicating "1.6" on a similar card, divided into twentieths*)?

Greg: One point one-two, or one point twelve.

Teacher: Right, OK. Now all I've done, these are all pointing at the same place. (*He lines the cards up one under another.*) Noughts are in the same place, all right?

Greg: Yeah.

Teacher: All I've done is chopped the scale into different ...

Greg: (*interrupting*) ... sections.

Teacher: Sections, right. And I've marked the same place on each scale. (*The arrows he put on the cards are clearly one under another.*) OK?

Greg: So they're all the same really.

Teacher: So what should they be?

Greg: Um ... (*puzzled by the question*)

Teacher: Do you think they're all right?

Greg: Yes, yes. (*confidently*) They're all right, but ... um ...

Teacher: So if I was to mark in an extra point there, and there, and there, and there (*altering one of the scales*), this would not be one point three anymore?

Greg: No, it'd be one point six.

Teacher: It'd be one point six, and if I put in extra points between those?

Greg: It'd be one point twelve.

Teacher: It'd be one point twelve, so it does matter which scale I use, does it?

Greg: Yes.

So what is going on? Decimals reappear and Greg is in trouble again, although it seems as if he thinks the problem is the teacher's. He sees no incongruity in the fact that different divisions on the scale have different names but are "the same really." The clue lies in his earlier language. It is clear that Greg approached the learning of decimals through the concepts of fractions ("nought point four tenths and five hundredths," "put in another half, a sort of point five"). His notion of decimals was firmly tied to, and drew on, his strong understanding of fractions. One of the activities Greg had engaged in when working on fractions was folding sheets of paper to physically create visual fractions of the whole sheet. One such activity involved him in folding the sheet in two to get halves and then opening it out and coloring one-half. He then refolded it in half and folded it again to get quarters and again to get eighths. When the sheet was unfolded, the creases in the portion that Greg had colored revealed that one-half is the same as two-quarters, which is the same as four-eighths. The size of the shaded piece of paper was the same, no matter how it was divided, but the names given to the pieces did depend on how it was divided. Although verbally different, *one-half, two-quarters,* and so on, are of course all correct, all "the same." If one can have equivalent fractions, then is it not reasonable to assume that under the redivision of a linear scale, *equivalent* decimals are possible? Certainly Greg thought so.

Listening to "Everything"

If careful listening can offer teachers clues to the thinking of the students, then it is natural to assume that the students, too, will be listening for clues to what the teacher expects of them. Many teachers are unaware that they give as much information away with their behavior and tone of voice as they do with their actual words. Students are not slow to pick up on this. When being questioned, they are often watching and listening to more than the teacher's words to help them fathom what answer is wanted. In the final classroom episode, we see two girls listening to the teacher's words, tone of voice, and body language as they struggle to reveal their understanding of decimals. Two words are problematic in this exchange: *bigger* and *number*. Neither word is overtly addressed by the teacher. Through the actions and words of the students, the videotape of the interaction reveals clearly their attempts to supply the required answer to a question that they only imperfectly understand. Listening to no more than the girls' words, however, the teacher sees the difficulty as one of mathematical understanding and not one of language.

The discussion begins with the teacher's writing the numbers *0.64* and *0.8* one above the other and asking the girls, "Which one is bigger?" Jane confidently replies, "Eight," interpreting the question to be concerned with the digits 0, 4, 6, and 8. The teacher's response, however, is not what she is expecting:

Teacher: No. Which number is bigger? This number or this number (*clearly underlining with his finger the 0.64 and the 0.8*)?

Jane: (*Confidently*) The top number (*underlining only the 64*).

Teacher: The top number is bigger (*stated, not posed as a question*). Why?

Susie: (*Reassured by the teacher's tone that the answer is correct*) Because sixty-four ...

Jane: Yeah, it's got more figures in it than that bottom number.

Teacher: Oh, I see. Do you think numbers with more figures in them are bigger, then? (*slight pause*) Always?

Jane: (*Hastily responding to the negative implication of the pause*) No, not always.

Teacher: Can you give me an example when a long number is smaller?

Jane: (*After a pause*) If you've got nought point one nought nought nought.

Teacher: (*Writing 0.1000 below the other two numbers*) Yes, and is that big or small compared with this one (*points at 0.8*)?

Susie: Small.

Teacher: Which is bigger, nought point eight or nought point one nought nought nought? Which is bigger?

Susie: (*Changing her response because she seems to think that "teachers don't ask the question again if you got it right" [interview data]*) The bottom one. That one (*pointing at 0.1000*).

Teacher: How can you tell? (*pause*) Which is the bigger of these two (*pointing at 0.8 and 0.1000*) and why?

Jane: That one's bigger (*pointing at 0.8*) because it's got an eight instead of a one … 'cause eight is higher than one.

What meaning are the girls building for "bigger" in this context? So far they have deduced that "more figures," paraphrased by the teacher as "a long number," does *not* mean "bigger." Listening for clues, we hear Jane say "higher," but her meaning in this context is not explored, although it is assumed that she is not referring just to the fact that it is written higher up the page. The lack of common meaning for *number* has gone unnoticed and recurs later in the lesson, still causing misunderstanding because for Jane *number* is synonymous with *whole number.*

Classroom communication has two inherent dangers that can be reduced by a teacher's listening sensitively. The first of these dangers results from a perceived imbalance of power, which can inhibit students from questioning the language used by the teacher. Jane and Sarah do not ask, "What do you mean by bigger?" when they realize that their understanding does not match that of the teacher. They watch and listen and try to guess at the meaning. The second danger is more universal: until the talk ceases to be compatible with the thinking of one of the participants, each participant will be assuming that all have a common understanding of the words being used. The students do not at this point challenge the teacher's meaning of the word *number,* nor he them. Neither has listened for this difference of meaning.

Later in the lesson, the girls' voice inflections and reactions reveal two things to the listening teacher: the students have discovered the possible meaning of *bigger,* and the background of their decimal understanding is very different from that of Greg.

Teacher: Let me show you something (*drawing a number line from 0 to 2 marked off in tenths*). Where's nought point eight on that number line?

Jane: There (*points to it with no hesitation*).

Teacher: We put an arrow and nought point eight. All right? Where's nought point six four?

Susie: About there (*points, with confidence, to between 0.6 and 0.7*).

Teacher: (*Unable to see clearly where she is pointing*) About?

Susie: Just before halfway between.

Teacher: Nought point six four (*marking on number line*). Which is bigger?

Susie: Nought point eight (*without hesitation*).

Jane: That one (*pointing, with total confidence, to 0.8*).

The students' siting of 0.8 and the very precise placing of 0.64 "just before halfway between" 0.6 and 0.7 shows clearly that they do have quite a developed understanding of decimals. As soon as the number line was produced, they were confident in their answers. It was almost as if they had said, "If only you'd told us that by *bigger* you meant further along the number line, we would have had no problems." One can see that Jane's original problem was one of language rather than of understanding.

CONCLUDING COMMENTS

It is hoped that through the study of the protocols presented in this article, readers have seen the possibilities for enhancing their practice. The intention has been to raise in the reader's mind questions about when and how to paraphrase a student's oral contributions, when it might be appropriate to introduce mathematical terminology (either casually or with a specific definition), when it might be appropriate to persuade students to attempt to reformulate their conceptions orally, and whether distracting information is passed to the student through a teacher's use of particular language and voice inflection. The reader's attention has been drawn to the need to attend to the subtle power of the individual words that students use; mathematics is about precision of thought, and this is best expressed through precision of language, be it verbal or symbolic.

The aim of this article has been to illustrate some of the insights that can be gained by consciously and specifically paying close attention to the oral communication of the students, in other words, by actively listening, not simply hearing what they say.

BIBLIOGRAPHY

Backhouse, John, Linda Haggarty, Susan Pirie, and Jude Stratton. *Improving the Learning of Mathematics.* London: Cassell, 1992.

Barnes, Douglas, James Britton, and Mike Torbe. *Language, the Learner and the School.* New York: Viking Penguin, 1986.

Durkin, Kevin, and Beatrice Shire. *Language in Mathematical Education: Research and Practice.* Milton Keynes: Open University Press, 1991.

Labinowicz, Ed. *Learning from Children.* Menlo Park, Calif.: Addison-Wesley Publishing Co., 1985.

Mathematical Association. *Maths Talk.* Cheltenham, England: Stanley Thornes, 1987.

Pimm, David. *Speaking Mathematically.* London: Routledge & Kegan Paul, 1987.

Pirie, Susan E. B., and Tom E. Kieren. "Growth in Mathematical Understanding: How Can We Characterise It and How Can We Represent It?" In *Learning Mathematics: Constructivist and Interactionist Theories of Mathematical Development,* edited by Paul Cobb, pp. 61–86. Dordrecht: Kluwer, 1994.

Wrangham, C. "Are You a Listening Teacher? Tape Yourself and See." *Modern English Teacher* 1 (1992): 25–27.

16

Developing Problem-Solving Behaviors by Assessing Communication in Cooperative Learning Groups

Alice F. Artzt

With the recent emphasis on providing opportunities for students to communicate about mathematics, cooperative learning strategies have taken on new significance. The small-group setting appears to provide a natural environment in which increased dialogue and communication about mathematics can occur among students. According to Patton, Giffin, and Patton (1989, p. 11), "Communication is the essence of the small-group experience."

Research has suggested, however, that for positive outcomes to occur, small-group activities must be structured to maximize the chances that students will engage in questioning, elaboration, explanation, and other verbalizations in which they can express their ideas and through which the group members can give and receive feedback (Slavin 1989). Without structure, when students are merely asked to work together with their group to solve a problem, complete an assignment, or just help one another as needed, the communication among students can take many forms. The students may just share answers or do one another's work, or they may help one another. The possible forms of communication can be represented on a continuum ranging from students who (although seated in a group) work independently and do not communicate with one another to students who all communicate with one another in the solution

Much gratitude goes to Frances R. Curcio and Eleanor Armour-Thomas for their help in the design of the instrument, the coding of the data, and the editing of the manuscript. Additional thanks go to Michele Uzbay and her class of students at Lawrence Middle School for the photograph.

of the problem. Between these two extremes are a multitude of possible communication patterns. Figure 16.1 (adapted from Artzt and Armour-Thomas [1992]) depicts several situations that can occur.

The most successful grouping strategies are those that are carefully structured so that a group goal fosters positive interdependence among the members and so that each person is held individually accountable for the work that is done in the group. This result can be accomplished through appropriately designed incentive structures and tasks (Artzt and Newman 1990). In addition, to ensure that students will engage in the face-to-face interaction required for communication, they must feel happy in their groups and they must have appropriate interpersonal and small-group social skills (Artzt 1994). Therefore, careful attention must be given to how the groups are formed and time must be allotted for group members to reflect on and assess the participatory and social skills of the group members (Johnson, Johnson, and Holubec 1986).

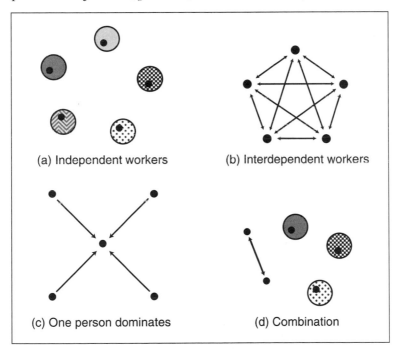

(a) Independent workers   (b) Interdependent workers

(c) One person dominates   (d) Combination

Fig. 16.1. Patterns of group interactions. Adapted from Artzt and Armour-Thomas (1992). Reprinted with permission.

Although a necessary component of effective group communication is the social and participatory behavior of the students, the most important aspect of the discourse is what it reveals about students' understanding of mathematics, or more specifically, what it reveals about the problem-solving behaviors of the students as they work within their groups. Such behaviors and understanding are difficult to assess, and only recently have they been addressed by researchers in the field. Artzt and Armour-Thomas (1992) attempted to address this issue through the development of a cognitive-metacognitive problem-solving framework for assessment. The theoretical framework as well as the means to adapt it to a classroom situation will be discussed next.

ASSESSING COMMUNICATION FROM A PROBLEM-SOLVING PERSPECTIVE

Given that the *Curriculum and Evaluation Standards for School Mathematics* (NCTM 1989) stresses problem solving and mathematical reasoning as important components of mathematical power, any assessment of discourse should focus on the individual and collective reasoning and problem-solving behaviors of the members of the group. The following is a description of a theoretical framework and an assessment instrument through which such assessment can be accomplished (Artzt and Armour-Thomas 1992; Curcio and Artzt 1995).

An Instrument to Assess Communication in Small Groups

In order to evaluate the reasoning and problem-solving behaviors of students during small-group work, the framework underlying the assessment instrument categorizes the statements that students make and the behaviors that students exhibit. This classification allows for the examination of both the mathematical heuristics and the cognitive processes students use as they engage in problem solving in a small-group setting. This framework expands on Polya's (1945) conception of mathematical problem solving as a four-phase heuristic process (i.e., understanding, planning, carrying out the plan, and looking back) by incorporating the types of problem-solving behaviors students exhibit as they solve a mathematical problem in a small group. These behaviors can be categorized as *reading, understanding, exploring, analyzing, planning, implementing, verifying,* and *watching and listening.*

Furthermore, each of these behaviors can be categorized as predominantly *cognitive* or predominantly *metacognitive,* where the working technical distinction between cognition and metacognition is that *cognition* is involved in doing, whereas *metacognition* is involved in choosing and planning what to do and monitoring and regulating what is being done. These categories of cognition are included because recent research has begun to identify the importance of metacognition in mathematical problem solving (Schoenfeld 1987). Artzt and Armour-Thomas (1992) have identified monitoring and self-regulation as metacognitive behaviors that are essential for successful problem solving in mathematics. Small problem-solving groups serve as natural settings for discussion in which the interpersonal monitoring and regulating of members' goal-directed behaviors occur. These factors may well be responsible for the positive effects observed in small-group mathematics problem solving, and so these are the factors that one would wish to encourage and assess as students participate in small mathematics-problem-solving groups. In fact, in their application of this framework, Artzt and Armour-Thomas (1992) and Curcio and Artzt (1995) found that a continuous interplay of cognitive and metacognitive behaviors is necessary for successful problem solving and maximum communication among students. Additionally, in successful, communicative groups, problem-solving behaviors occur intermittently with students' returning several times to read, understand, explore, analyze, plan, implement, or verify. These behaviors mirror the self-communication behaviors of the expert problem solver working alone, which was documented by Schoenfeld (1987). (Researchers who are interested in a detailed description of the framework should see Artzt and Armour-Thomas [1992].) The instrument has been adapted for use by classroom teachers and students alike. A description of this assessment instrument and how it can be applied follows.

Application of the Instrument

As students work in a small group to solve a mathematical problem, their behaviors can be categorized according to the chart in figure 16.2. To use the instrument, the teacher could observe a group and record the behaviors on the chart under four broad categories: Talking about the Problem, Doing the Problem, Watching or Listening, and Off Task. The sample chart in figure 16.2 contains descriptors within each category for use as a checklist to monitor one group working on a problem each day. The teacher systematically examines the behaviors of each person for approximately one minute in order to note the proper category or categories. For example, if a group of four students worked for twelve minutes, at least three recordings of the behaviors of each student would be made. Space is left at the bottom of the chart to include important anecdotal information that further describes the students' behaviors. For example, the teacher might wish to point out which students were responsible for making what Schoenfeld (1985) calls "executive decisions." These are the key decisions that can account for the success or failure of the work. They may consist of insightful plans for how to solve a problem, or they may be acute observations that the problem solution is going in the wrong direction.

To see a specific example of how this instrument can be applied, let us examine the following excerpt from a group discussion that was recorded from a group of seventh-grade students who were trying to solve the following problem:

> A banker must make change of $1.00 using 50 coins. She must use at least 1 penny, 1 nickel, 1 dime, and 1 quarter. What combination of coins does she use?

	Student A	Student B	Student C	Student D
Talking about the Problem				
Understanding				
Analyzing				
Planning				
Exploring				
Implementing				
Verifying				
Doing the Problem				
Reading				
Exploring				
Implementing				
Verifying				
Watching and Listening				
Off Task				
Anecdotal Information:				

Fig. 16.2. Chart for categorizing student behaviors using problem-solving descriptors

This problem is especially suitable for several reasons. First, since no algorithm is available to students at this level that would enable them to solve the problem in a straightforward manner, the students must engage in a guess-and-test approach. This approach affords students the opportunity to practice organizing information systematically, keeping track of units, and working with decimals. Second, since a guess-and-test approach is needed to solve this problem, it conforms to what Cohen (1994, p. 8) describes as an "ill-structured problem." According to Cohen, the successful solution of ill-structured problems depends on the degree of communication within the group. In short, the successful solution of this particular problem requires maximum communication among students. To help highlight the different levels of behavior that can occur in a group, the students selected for the group display very different levels of problem-solving ability.

Because the coding of behavior is based on what the teacher sees as well as hears, an overview scenario is presented of the behaviors and the statements of four students (Anna, Brian, Conny, and Domingo) during the last five minutes of a fifteen-minute problem-solving session. This scenario is followed by the coded protocol of the specific statements and behaviors of these four students. Note that the coding decisions are made on the basis of the salient behaviors of the students rather than on the basis of each individual statement.

Overview

Anna began with a *plan* about how to approach the problem but overlooked the requirement to use at least one penny. Anna's strategy apparently gave Brian the idea of what to do; however, he did it correctly, using the penny. Brian *analyzed* the problem aloud but did not pick up a pencil to implement his ideas. Anna did the calculations, and using Brian's ideas, she *implemented* the plans and succeeded in solving the problem. Conny and Domingo were trying to follow the problem-solving process but were unable to keep up with the ideas of Anna and Brian. At this late stage in the problem-solving process, they both seemed still to be struggling with trying to *understand* the problem. Anna and Brian made no effort to try to help them to understand. Judging from the discourse, we could assume that Anna, who solved the problem, and possibly Brian understood how to solve the problem. It is important to note as well that Anna might not have solved the problem without having listened to Brian, who put her on the right track.

(*Planning*)	Anna:	Why don't we just use everything. A dime, a quarter, and a nickel. How much is that?
(*Implementing*)	Domingo:	A dime, a quarter, and a nickel? That's 40.
	Brian:	41 cents.
	Anna:	OK, 40 cents. OK, we have 3 coins.
	Domingo:	I still don't understand what you're talking about.

	Brian:	Listen to this. I have an idea. If we use...(*interrupted by Conny*)
(*Understanding*)	*Conny:*	What about the 50?
(*Analyzing*)	*Brian:*	If we use every coin so far only once, we'll have 41 cents, and we'll have 46 coins left over.
	Anna:	What are you doing?
	Brian:	We used every coin so far, so we don't have to worry about it any more. So we have 41 cents, and we have 46 coins to use. We have to use more pennies.
(*Understanding*)	*Conny:*	How much does it have to be?
	Brian:	It has to be a dollar.
(*Implementing*)	*Anna:*	OK. So you have 4 coins. What if we add 46 pennies?
	Conny:	(*Adding it up*) That's 98.
(*Understanding*)	*Brian:*	Oh, wait. We have to use all 50 coins.
	Anna:	That's what I keep telling you.
(*Planning, Analyzing*)	*Brian:*	Let's do what we did before—a quarter, a dime, and a nickel and a penny. That's 41 cents, and we used 4 coins.
(*Implementing*)	*Anna:*	(*After calculating and working with her idea of 46 pennies*) I got it!
(*Analyzing*)	*Brian:*	We need 46 more coins.
(*Verifying*)	*Anna:*	I got it. Look at this. Five coins. (*Everyone looks at her paper as she verifies the solution in writing.*)
	Domingo:	Yeah, it works—45 pennies. It works. Yes.

Coded behaviors and statements

Figure 16.3 displays the completed chart that represents the communication in this group. Examining this chart, we can see that Anna and Brian were responsible for most of the mathematical communication that transpired in the group. They engaged in the majority of the problem-solving behaviors that led to the problem solution. Conny and Domingo were on task throughout the session, but their roles were clearly more passive, and they did only a few calculations that were suggested by the others. Unlike Domingo, Conny showed some sign of understanding what the problem was about. The example shown displays only a small piece of a group's work. To increase the involvement and ensure Conny's and Domingo's understanding, provisions for individual accountability would follow. That is, group tasks would be designed in which the completion of the group's product depended on every student's unique contribution to the work.

At the completion of cooperative learning activities, it is important to conduct group processing sessions in which students are called on to assess their participatory and social skills in the group. At this time students reflect on and discuss how well they have worked individually and together.

	Anna	Brian	Conny	Domingo
Talking about the Problem				
Understanding	✓	✓ ✓	✓ ✓	
Analyzing		✓ ✓ ✓ ✓		
Planning	✓	✓ ✓		
Exploring				
Implementing	✓ ✓			
Verifying				
Doing the Problem				
Reading				
Exploring				
Implementing	✓	✓	✓	✓
Verifying	✓			
Watching and Listening	✓	✓ ✓ ✓ ✓	✓ ✓ ✓ ✓	✓ ✓ ✓
Off Task				

Anecdotal Information:

Anna came up with a good plan with a minor fault in it. Brian correctly refined Anna's plan, and then Anna did most of the work on her own to solve the problem. Brian kept analyzing and exploring the problem out loud but never made any calculations himself and did not follow what Anna was doing. Conny and Domingo mostly looked on, trying, unsuccessfully, to understand what was going on. Anna and Brian made little effort to communicate their understanding to Conny and Domingo.

Fig. 16.3. Sample chart of coded behaviors

By using the foregoing framework to document the group's communication during the group processing session, the teacher and students can focus on the individual problem-solving behaviors, including the monitoring and regulating behaviors that contribute to the successful solution of problems. To further increase the students' awareness of the role of different problem-solving behaviors, the instrument can be used by the students themselves. That is, they can take turns assessing the problem-solving behaviors that occur in their own group, or they might even observe other groups. Most important, by becoming aware of the significant role of such problem-solving behaviors as *reading, understanding, analyzing, planning, exploring, implementing,* and *verifying* and of the importance of monitoring and regulating a problem solution, students can begin to try to incorporate these types of behaviors into their individual problem-solving efforts.

They can also come to understand that the use of these behaviors does not necessarily occur in a linear progression; that is, students can return several times to different behaviors. A problem may be read several times throughout the course of the problem solving. Rereading allows students to gain a deeper understanding of the problem's requirements through the course of the problem-solving session. A plan may be incorrect, leading to a verification process that points out the mistake, which may lead the group back to a new analysis, an exploration, or a new plan. It is important for students to recognize this process so that when they are engaged in individual problem solving, they don't quit the minute a plan doesn't work. Most important, they must come to understand the value of monitoring and regulating their work during problem solving. That is, they must realize the value of asking one another such questions as "What are you doing?" "Why are you doing that?" "How will that help us solve the problem?" They must be encouraged to transfer this type of communication that they engage in with their group members to communication they have with themselves when they work alone. As Schoenfeld (1987, p. 210) suggests, good problem solvers are "people who are good at arguing with themselves."

CONCLUDING COMMENTS

The importance of communication in learning mathematics has deservedly been given much attention by the community of mathematics educators. Of utmost importance is the nature of this communication. Through the use of effective cooperative learning and an assessment scheme that values the communicative behaviors that are associated with effective problem solving, students can learn to engage actively in mathematical reasoning in a social setting. The cognitive processes students use when interacting with their group members can then be internalized and activated when they work individually (Bershon 1992). In effect, the communication they encounter in their cooperative learning groups becomes the communication they have with themselves when working alone.

REFERENCES

Artzt, Alice F. "Integrating Writing and Cooperative Learning in the Mathematics Class." *Mathematics Teacher* 87 (February 1994): 80–85.

Artzt, Alice F., and Eleanor Armour-Thomas. "Development of a Cognitive-Metacognitive Framework for Protocol Analysis of Mathematical Problem Solving in Small Groups." *Cognition and Instruction* 9 (1992): 137–75.

Artzt, Alice F., and Claire M. Newman. *How to Use Cooperative Learning in the Mathematics Class.* Reston, Va.: National Council of Teachers of Mathematics, 1990.

Bershon, Barbara L. "Creative Problem Solving: A Link to Inner Speech." In *Interaction in Cooperative Groups: The Theoretical Anatomy of Group Learning,* edited by Rachel Hertz-Lazarowitz and Norman Miller, pp. 36–48. New York: Cambridge University Press, 1992.

Cohen, Elizabeth. "Restructuring the Classroom: Conditions for Productive Small Groups." *Review of Educational Research* 64 (Spring 1994): 1–35.

Curcio, Frances R., and Alice F. Artzt. "Students Communicating in Small Groups: Making Sense of Data in Graphical Form." Unpublished manuscript, 1995.

Johnson, David W., Roger T. Johnson, and Edythe Johnson Holubec. *Revised Circles of Learning: Cooperation in the Classroom.* Edina, Minn.: Interaction Book Co., 1986.

National Council of Teachers of Mathematics. *Curriculum and Evaluation Standards for School Mathematics.* Reston, Va.: National Council of Teachers of Mathematics, 1989.

Patton, Bobby R., Kim Giffin, and Eleanor N. Patton. *Decision-Making: Group Interaction.* New York: Harper & Row, 1989.

Polya, George. *How to Solve It.* Garden City, N.Y.: Doubleday, 1945.

Schoenfeld, Alan H. *Mathematical Problem Solving.* Orlando, Fla.. Academic Press, 1985.

———. "What's All the Fuss about Metacognition?" In *Cognitive Science and Mathematics Education,* edited by Alan H. Schoenfeld, pp. 189–215. Hillsdale, N.J.: Lawrence Erlbaum Associates, 1987.

Slavin, Robert. *School and Classroom Organization.* Hillsdale, N.J.: Lawrence Erlbaum Associates, 1989.

17

Using Technology to Enhance Communication in Mathematics

Ann E. Barron

Michael C. Hynes

The students in Ms. Kruger's middle school mathematics class had just finished viewing The Adventures of Jasper Woodbury: Journey to Cedar Creek. The discussions were very lively as they broke into groups to try to help Jasper figure out if he could make it back home before sunset.

The students quickly came to the conclusion that they needed to know whether Jasper had enough time and enough fuel to get home. But, wait—having enough fuel depended on the capacity of the fuel tank and the amount of money available to buy more fuel. The students also realized that the current of the river might be a factor in the solution. Wow! This problem was complicated and really made them stop and think. It involved observation skills, deductive reasoning, map-reading ability, and conversion of various data types.

Ms. Kruger smiled to herself as she watched and listened to the discourse taking place in the groups. One group was using the bar-code reader to review pertinent information from the videodisc to help them estimate distances along the river, another was discussing the shape of fuel tanks and estimating procedures, a third group was creating a scale representation of the map with a graphics program, and a fourth group was debating about the importance of the river current in the problem. Ms. Kruger noted that a hands-on laboratory on the relationship between the volume of a container and its capacity might be a valuable follow-up lesson for the students. Also, she decided to introduce the notion of vectors to see if any of the students would make the connection to the discussion of river currents.

The students were motivated and interested. They used calculators for complex calculations and employed a variety of software programs to investigate the mathematical problems and subproblems involved in the real-world scenario. They knew that the emphasis was not on arriving at a single "right" answer. Instead, they would be required to present their approach and conclusions orally to the rest of the class. The visual defense of their findings would be conveyed in a hypermedia program that contained the graphs, charts, and other information they developed to represent the concepts. If they chose, the hypermedia program could also become a part of their yearly assessment portfolios.

THE underlying value of the richness of the Jasper Woodbury problem, presented through technology, is the opportunity for students to formulate their own questions, to clarify the problem, to communicate their understanding of the problem in their group, to come to a consensus about a way to solve the problem, and to plan a presentation of their solution to the class. This communication of mathematical ideas empowers students as learners. The teacher's role becomes facilitating and encouraging meaningful discourse among members of the group. The teacher also should identify areas of student interest sparked by the problem that may lead to more rich mathematical experiences. In this article we will discuss various types of computing technology designed to enhance communication in mathematics.

This use of videodisc-based problem solving is but one of many ways in which technology can foster communication in a mathematics classroom. This paper examines the use of software, networks, multimedia, and programming to enhance communication in mathematics.

CLASSROOM USES OF GENERAL-APPLICATIONS SOFTWARE

General-applications software includes word processing, spreadsheets, databases, and graphics programs. With these tools, students can communicate with and about mathematics. Word processors empower students to express their mathematical ideas in writing; spreadsheets can help free students from the tedium of repetitive mathematical calculations; data comparisons are possible with database-management programs; and students can use graphics programs to create drawings to enhance solutions. All these tools can help students engage in authentic learning activities (see fig. 17.1). For example, in a unit on money and banking, students can use databases to keep records and generate reports, graphics programs to print currency, telecommunications to track stock prices, and spreadsheets to create charts (Klenow 1993).

Desktop Adventures

Desktop adventures are another example of using applications software for problem solving. In a desktop adventure, an integrated software program (one that contains at least a word processor, a spreadsheet, and a database) is used to create an environment in which students can easily navigate from one program to another to obtain the necessary information. For example, the word-processing component might contain the narrative of the problem, perhaps setting the stage with clues to solve a mystery. The answers to the clues, however, would be located in the spreadsheet or database. The idea is for students to have all available applications open simultaneously to access, record, organize, and interpret the data as necessary.

Word Processors

- Write solutions to problems
- Formulate and write student-generated problems
- Express mathematical ideas in writing
- Realize that reading and writing mathematics are a vital part of learning and using mathematics

Spreadsheets

- Relate charts to mathematical ideas
- Organize data
- Make conjectures
- Write mathematical expressions
- Construct formulas
- Clarify thinking about number patterns
- Analyze data
- Select appropriate forms of graphs and charts
- Explore recursive change

Databases

- Formulate pertinent problems or questions
- Identify data attributes that are germane to a problem
- Organize data
- Interpret quantitative data
- Express mathematical ideas through reports

Graphics Programs

- Relate pictures and diagrams to mathematical ideas
- Model mathematical situations using graphics
- Use graphics to clarify thinking about mathematical ideas and relationships
- Enhance written mathematical presentations with graphics
- Make conjectures

Fig. 17.1. Learning activities that can be achieved through applications software

Using a desktop-adventure approach, Mr. Hernandez challenged his fifth-grade students to determine the important factors in pizza sales. A word-processing file introduced the problem, and integrated spreadsheets were designed to help students explore the area and circumference of circles, to show the cost of buying more than one pizza when the price varies, and to examine the profits and losses in pizza sales for a fictitious school cafeteria. Students had to decide which spreadsheets contained pertinent information about the problem and use the data appropriately. The integrated database application included the menu information from several fictitious pizza shops. The sizes, shapes, and costs of pizzas were also available. The phone numbers and addresses of several local pizza shops were located in another database. Thus, students could collect real data from their community or use fictitious data to investigate the problem and formulate a solution.

The desktop adventure presented in the example is rich in meaningful mathematics. Students can explore the meaning of exponents as they relate to the area of a circle. How do you describe a circle? Does a twelve-inch pizza have a twelve-inch radius, diameter, or circumference? The decision the students make will determine the nature of their arguments. Since Mr. Hernandez teaches in an interdisciplinary team, the social studies teacher can have students explore truth in advertising while this problem is being considered in mathematics.

Assessment Portfolios

An assessment portfolio is a deliberate collection of a student's work and accomplishments. It is usually compiled by the student and the teacher in collaboration. With the changing emphasis in the assessment of mathematics learning, application software can be an important tool because students can use spreadsheets, databases, and other software in the development of their portfolios.

Many states and individual school districts require portfolios as a part of the student-assessment program (Saylor and Overton 1993). In Kentucky, all students keep mathematics portfolios and writing portfolios. The mathematics portfolios are designed to illustrate the students' accomplishments throughout the year and to portray a philosophy that—

- focuses on strengths rather than weaknesses;
- values a variety of learning styles;
- values mathematics as a subject that requires careful and thoughtful investigation;
- encourages students to communicate their understandings of mathematics at a high level of proficiency;
- promotes a vision of mathematics that goes beyond correct answers;
- emphasizes the role of the student as the active mathematician and the teacher as a guide.

One problem with using portfolios is the storage and retrieval of the information. Technology can be helpful. Word processors, spreadsheets, and databases allow students to store their presentations, answers to performance items, and so on, on diskette. If scanners are available, even graphics associated with their work can be included. Some schools are experimenting with more-advanced technology that stores the written materials, photographs, and video recordings on videodiscs or compact discs.

NETWORKS

Computer networks are becoming very popular in schools. Networks connect several computers and allow students to communicate within and beyond the classroom. The most common forms of networks used for education are local area networks and telecommunications. Each of these networks can enhance communication in the mathematics class.

Local Area Networks

A local area network (LAN) is a group of computers that are physically connected, usually through wires. Each computer has a special card that allows it to communicate with other computers in the same network. Many schools are networking computers because it allows expensive peripherals, such as printers and CD-ROM players, to be shared. It is also less expensive to purchase one LAN copy of a software program and install it on a central file server for all computers to access than to purchase individual copies of software.

When students use computer software on a LAN, they can share software programs and files in the program. For example, several students could be working with the same database program, and the LAN software would ensure that each student has access to the latest updates. With appropriate software, a teacher can project one computer display on all the other screens. For example, if a teacher were teaching a new graphics program, he or she could display the instructor's screen on all the students' computers and the students could follow the screen changes on their individual monitors. Likewise, if the students were studying the Pythagorean theorem using an application such as Geometer's Sketchpad, the teacher could project an individual student's screen, showing her or his explorations into the relationship of the hypotenuse and the legs of a right triangle, onto all the other students' screens.

Several software programs are designed for student collaboration on a LAN. For example, the network version of the Oregon Trail allows several students to travel "together" to Oregon. Each student works at an individual computer, but the decisions made by one student will affect all the others. Through a LAN, students can share information, communicate with one another, send electronic messages to the teachers, and interact on a real-time basis.

Telecommunications

Telecommunications offers wonderful opportunities for mathematics students and teachers to communicate with peers and experts beyond their classrooms. The basic components necessary for telecommunications consist of a computer, a modem, a telephone line, and software. These components are generally quite inexpensive because the computer can be a very basic model and telecommunications software can be obtained as shareware. As LANs become more common in schools, many districts are connecting the school LAN directly to a larger network, such as the Internet.

Telecommunications provides endless communication and research opportunities for mathematics education. Some of the instructional activities involve a simple exchange or transfer of information; others involve high-level problem solving and synthesis. Figure 17.2 outlines some of the activities that can be integrated into the mathematics class with a simple e-mail connection through a statewide educational network, commercial telecommunications services (such as Prodigy), a local bulletin-board system, or the Internet.

Turtle Trips
Students interact with another classroom to send Logo commands and create Logo pictures and programs.

Mathematical Mysteries
Students create logic problems, puzzles, and other mathematical mysteries for another classroom to analyze and solve.

Pooled Data Analysis
Students conduct surveys, collect data, and compare data (such as food prices, rainfall, etc.) from around the world. Internet sites, such as Weather Underground, are great places to obtain up-to-date information that can be analyzed, graphed, and used for predictions.

Vacation Planning
Students work in small groups to choose a site they would like to visit, and they contact some peers in that location. Plans to visit the location include the best routes, travel costs, sites to see, daily expenses, and so forth. Charts and graphs can be created comparing the costs of alternative locations and modes of travel.

Fig. 17.2. Sample telecommunications activities for mathematics class

The Internet

The Internet is a worldwide, high-speed network that connects thousands of other, smaller computer networks. If a school has access to the Internet, the students can interact with others all over the world and engage in pen-pal activities, peer-to-peer tutoring, and interactions with experts.

Students can also access remote computers to conduct searches, transfer files, and obtain information. Thousands of sites throughout the world allow free, anonymous access. The following are a few of the sites that are especially relevant for mathematics:

- The NCTM home page (http://www.nctm.org) has been designed to offer easy access to information about NCTM conventions, publications, and other activities. A job hotline, indexes of articles, selected student-ready sheets, "news briefs," and more are being discussed. Ideas are welcome from members about services they would find useful. Write to htunis@nctm.org with your suggestions.
- NASA Spacelink (http://spacelink.msfc.nasa.gov or telnet to spacelink.msfc.nasa.gov) is designed specifically for K–12 students and teachers. It contains databases with "historical documents, scientific information, current events, down-loadable graphics and interactive communication opportunities, in addition to classroom lessons and other teacher materials" (Kelly 1994, p. 178).

- Weather Underground (telnet to madlab.sprl.umich.edu) offers current and long-range weather for regions and cities. It also gives severe-weather advisories, ski conditions, and hurricane and earthquake reports. The information available at this and similar sites provides excellent data for analyses, investigations of trends, and predictions based on past facts.
- NEWTON (telnet to newton.dep.anl.gov) is Argonne National Laboratory's electronic bulletin board. The teachers' menus contain many discussions among science and mathematics teachers and students in all major curriculum areas. The AskA-Scientist option allows students to pose questions that will be answered in about three days. Discussion questions are given for ten general topics, one of which is mathematics.

Until 1994, most of the communications through the Internet consisted of cryptic, command-line instructions, such as >telnet spacelink.msfc.nasa.gov. The World Wide Web (WWW) now offers an environment in which documents, graphics, sound, and digital video can be linked and accessed with the click of a mouse button. Students can view on-line weather maps, interact with geometric shapes and figures, and generate charts from data located at distant sites.

Jamie Steward is a student in Ms. Magruder's class in upper elementary school. She became fascinated with Ms. Magruder's introduction of vectors. She began to think about the river near her grandfather's house. Sometimes the river current seemed swift, and at other points on the river it seemed slow. What made these differences? Why were there rough spots in the currents? There were so many questions.

Ms. Magruder told Jamie that she was beginning to think about a very important scientific question that involved mathematics. The field of study that considers such questions is computational fluid dynamics. She suggested that Jamie "surf the net" to see if she could find some additional information.

Jamie spent some time searching for a site that would have the information she needed. She struck gold on NASA Spacelink. In response to a question about computational fluid dynamics, she was given the name of a scientist at Ames Research Center at Moffett Field, Redwood City, California. Dr. Lucy Morse-Roberts informed Jamie that Ames was the center for this type of scientific research. Ames has many wind tunnels to study the nature of the flow of fluids across surfaces, and Dr. Roberts studies the phenomena using computer simulations. Jamie and Dr. Roberts have begun a long-distance e-mail conversation that may encourage Jamie to enter some field of scientific research.

COMMERCIAL SOFTWARE AND MULTIMEDIA

Computer software in itself does not promote the communication of mathematical ideas. However, software can be an effective tool in increasing the communication of mathematical understanding if the technology is used thoughtfully.

Commercial Software

To be compatible with the emphasis in the NCTM *Curriculum and Evaluation Standards* (1989) on understanding mathematics, software for teaching mathematics should be compelling, user-friendly, and highly interactive, and it should employ the power of the computer to create a learning situation that is different from that of worksheets or other teaching materials. Look for problem-solving software that has multiple solutions and encourages users to ask "What if?" questions. Examples of geometry software that encourages questioning are Geometric Supposer, Cabri, and the Geometer's Sketchpad. Students using programs like these can become actively involved in communicating in mathematical terms. They can also conjecture about, classify, and explore the concept of triangles. These activities can help them to make mathematical connections, thereby creating meaningful situations for cognitive processing (Brown, Collins, and Duguid 1989). Additional software programs that are designed for mathematics are included in figure 17.3.

Multimedia Materials

Multimedia instruction can be loosely defined as instruction integrating some, but not necessarily all, of the following media in an interactive environment that is controlled by a computer: text, graphics, animation, sound, and video. With the use of multimedia, students can use multiple modes of communication to understand mathematical ideas and to communicate their understanding to one another and to the teacher.

Commercial multimedia materials can encourage students to investigate problems, examine factors important to the problem, formulate hypotheses, solve problems, and present their solutions. Examples of software that provide students with rich problem-solving experiences involving many areas of mathematics include the following: Windows on Math (primary grades), Fizz and Martina (middle grades in elementary school), and The Adventures of Jasper Woodbury and Math Sleuths (middle school).

PROGRAMMING AND AUTHORING

Students also use communication skills when they work with computer languages. A host of communication skills are practiced each time a small program is written to solve a problem. Precise communication occurs as a student writes lines of code to solve the given problem. Students interpret error messages and rewrite the code. Comment lines are written to convey the purpose of blocks of code to another reader. Finally, students prepare and deliver demonstrations of programs they have written.

Graphing and Probability
  Graph Links (Harcourt Brace & Co.), Graphing and Probability Workshop
  (Scott, Foresman & Co.)

Measurement, Time, and Money Concepts and Skills
  Exploring Measurement, Time and Money (IBM K–12 Education)

Geometry Explored through Graphic Presentations
  Geometer's Sketchpad (Key Curriculum Press), Geometric Supposer
  (Sunburst), GeoDraw (IBM K–12 Education), Geometry Inventor (Sun-
  burst), Geoworks (Geoworks Co.), Cabri (Brooks/Cole)

Symbol Manipulation and Graphing
  Algebra Series (IBM K–12 Education), Derive (Soft Warehouse), Mathcad
  (MathSoft), Mathematica (Wolfram Research, Inc.), Toolkit for Interactive
  Mathematics (IBM K–12 Education), Maple (Brooks/Cole)

Order of Operations
  How the West Was One + Three x Four (Sunburst)

Ratio and Proportion
  My Travels with Gulliver (Sunburst)

Number Sense and Patterns
  Muppet Math (Sunburst)

Patterns, Problem Solving, Open-Ended Mathematics Exploration
  The Factory (Sunburst), Math and More (IBM K–12 Education)

Number Sense
  Exploring Math Concepts (IBM K–12 Education), Millie's Math House (EdMark)

Reference Software: Relates to Money, Computation, Estimation
  PC Globe (Broderbund)

Fig. 17.3. Selected commercial software for mathematics

Early in the development of computer languages, only a few people had
the technical skills to write programs because the languages were cumber-
some and tedious. The evolution of programming languages into high-
level, structured languages makes their use more valuable in school
mathematics classes. Ada, APL, C, Lisp, Pascal, BASIC, and many other
computer languages allow middle school and high school students to explore such mathematics concepts as variables, functions, and recursion
through programming activities. The introduction of the Logo computer
language gave even young children an opportunity to construct mathematics meaning from a programming language.

When Logo was introduced, it was anticipated that it would increase
students' problem-solving abilities and higher-order-thinking skills.
After nearly twenty years, the empirical evidence that such a benefit ex-
ists is contradictory. The conflicting results may be due to inappropriate

research designs, lack of random assignment to groups, lack of control groups, and incomplete documentation of the studies (Allocco et al. 1992; Ginter and Williamson 1985; Maddux 1989; Clements 1985). Despite the controversy, many proponents of Logo believe that the program helps students explore and discover mathematical relationships. Schools throughout the nation are designing unique and creative ways to incorporate Logo programs into mathematics classrooms (Taylor 1991; Bradley 1993; Bradley 1992).

In recent years, additional programming environments have been developed with hypermedia programs and icon-oriented design environments, such as LogoWriter, Logo MicroWorld, HyperCard, HyperStudio, LinkWay, and ToolBook, as well as with simulation-and-modeling languages, such as STELLA (Steed 1992) and TUTSIM (Klee 1994). The hypermedia programs present objectlike environments in which students can create interactive presentations and programs that integrate text, audio, graphics, and video (Bledsoe 1993). Recent research that analyzes programming in this new context "has shown a more positive relationship between programming and problem-solving" (Wiburg and Carter 1994).

CONCLUDING COMMENTS

In the fast-approaching twenty-first century, our students will experience a school environment that is drastically different from that of their parents and teachers. The integration of technology into school curricula is no longer a luxury; it is a prerequisite to survival in a future that will be driven and supported by technology (Mageau and Chion-Kenney 1994).

The challenge for teachers is to look for ways to make technology enhance the quality of communication about mathematical concepts, that is, to make mathematics learning more authentic. Teachers can take advantage of the potential of technology only if hardware and software are made available to *all* students, if teachers are prepared to use technology effectively, and if the school administration supports the use of technology in mathematics instruction. When these conditions exist, we can expect that meaningful mathematical discourse among students as well as between teachers and students will occur through the use of technology in mathematics instruction.

REFERENCES

Allocco, Lisa, Joan Coffey, Ann Marie Dalton, Justine Dariano, Joseph E. Dioguardi, Linda Galterio, and Brian Monahan. "To Teach or Not to Teach Logo: Reflecting on Logo's Use as a Problem-Solving Tool." *Educational Technology* 32 (August 1992): 23–27.

Bledsoe, Glen L. "HyperCard in the Math Classroom." *Computing Teacher* 21 (August/September 1993): 51–52.

Bradley, Claudette. "The Four Directions Indian Beadwork Design with Logo." *Arithmetic Teacher* 39 (May 1992): 46–49.

———. "Making a Navajo Blanket Design with Logo." *Arithmetic Teacher* 40 (May 1993): 520–23.

Brown, John S., Allen Collins, and Paul Duguid. "Situated Cognition and the Culture of Learning." *Educational Researcher* 18 (January/February 1989): 32–42.

Clements, Douglas H. "Logo Programming: Can It Change How Children Think?" *Electronic Learning* 4 (January 1985): 74–75.

Ginter, Dean W., and James D. Williamson. "Learning Logo: What Is Really Learned?" *Computers in Schools* 2 (Summer/Fall 1985): 29–31.

Kelly, M. G. (Peggy), and James H. Wiebe. "Mining Mathematics on the Internet." *Arithmetic Teacher* 41 (January 1994): 276–81.

Klee, Harold. "Systems Modeling." Course description, University of Central Florida, College of Engineering, 1994.

Klenow, Carol. "Count on Technology for Teaching Math." *Instructor* 102 (April 1993): 82–83.

Maddux, Cleborne. "Logo: Scientific Dedication or Religious Fanaticism in the 1990's?" *Educational Technology* 29 (February 1989): 18–23.

Mageau, Teresa, and Linda Chion-Kenney. "Facing the Future." *Electronic Learning* (October 1994): 37–40.

National Council of Teachers of Mathematics. *Curriculum and Evaluation Standards for School Mathematics.* Reston, Va.: National Council of Teachers of Mathematics, 1989.

Saylor, Kim, and June Overton. "Kentucky Written and Math Portfolios." Paper presented at the National Conference on Creating the Quality School, Middlesboro, Ky., March 1993. (ERIC Document Reproduction No. ED361382.)

Steed, Marlo. "STELLA, a Simulation Construction Kit: Cognitive Process and Educational Implications." *Journal of Computers in Mathematics and Science Teaching* 11 (Spring 1992): 39–52.

Taylor, Lyn. "Activities to Introduce Your Class to Logo." *Arithmetic Teacher* 39 (November 1991): 52–54.

Wiburg, Karin, and Bruce Carter. "Thinking with Computers." *Computing Teacher* 22 (October 1994): 7–10.

18

The Role of Open-Ended Tasks and Holistic Scoring Rubrics:
Assessing Students' Mathematical Reasoning and Communication

Jinfa Cai

Suzanne Lane

Mary S. Jakabcsin

PROPONENTS of the reform in mathematics education stress the need for students to use various modes of communication, such as written, oral, and visual expressions and physical manipulatives, to express their mathematical ideas.

This paper discusses the role of open-ended tasks and holistic scoring rubrics in assessing students' mathematical communication and reasoning. Open-ended tasks afford students opportunities to display their mathematical thinking, reasoning, and problem solving. Holistic scoring rubrics allow the criteria for evaluating students' responses to the open-ended tasks to be transparent to both teachers and students. In particular, by using holistic scoring rubrics, teachers can convey to students the criteria for evaluating their mathematical thinking so that students understand the differences between various levels of mathematical proficiency. Thus, the use of open-ended tasks and holistic scoring rubrics can facilitate teacher-student interaction and communication in classrooms.

Preparation of this paper was supported in part by a grant from the Ford Foundation (grant no. 890-0572) for the QUASAR project. Any opinions expressed herein are those of the authors and do not necessarily represent the views of the Ford Foundation.

THE USE OF OPEN-ENDED TASKS

In contrast to standardized multiple-choice items, one distinct feature of open-ended assessment tasks is that students are asked not only to produce their answers but also to show their solution processes and give justifications for their answers. Thus, open-ended tasks permit the display of students' mathematical thinking, reasoning, and communication. The role of open-ended tasks in assessing students' mathematical communication and reasoning can be clearly revealed by examining the multiple-choice task shown in figure 18.1 and its open-ended counterpart. Both versions are intended to assess students' conceptual understanding of "decimal place value," the value of places to the right of the decimal point. (The examples of open-ended tasks used in this paper are drawn from the QUASAR Cognitive Assessment Instrument [QCAI] [Lane 1993]. QUASAR [Quantitative Understanding: Amplifying Student Achievement and Reasoning] is a national project designed to improve mathematics instruction for students attending middle schools [grades 6–8] in economically disadvantaged communities [Silver 1993]. The QUASAR Cognitive Assessment Instrument is designed to monitor and evaluate the impact of programs.)

At first glance, the two tasks appear very similar, but because students are asked to explain their answers in the open-ended format, a seemingly ordinary, would-be multiple-choice item becomes a rich and informative task that brings mathematical communication into assessment. Students' written explanations for the open-ended decimal task reveal various bases for the justifications (Magone, Cai, Silver, and Wang 1994). For example, students' explanations may display their knowledge of decimal place value, of relationships between the decimal numbers and the decimal point, or of fractional equivalents of the decimal numbers.

Figure 18.1 shows three middle school students' responses to the open-ended decimal task. In the first response, the student's explanation suggests that the student transformed the decimals into equivalent fractions to explain why .8 is the greatest number among the four given decimal numbers. In the second response, although the student correctly selected .8 as the number with the greatest value, the rationale for the selection of .8 is incorrect. More important, this response shows the power of the open-ended task in assessing students' understanding of decimal place value. The open-ended task allowed students to communicate their understanding and therefore allowed a more comprehensive assessment of students' understanding of place value. In the multiple-choice version of the task, if a student chooses .8, it is assumed that the student understands decimal place value as it is assessed by the task. However, the second response to the open-ended task already shows that the student who chose .8 as the greatest number did not necessarily have sufficient knowledge to support the selection. The third response shows a student's misconception about decimal place value. This student used a "whole-number rule"

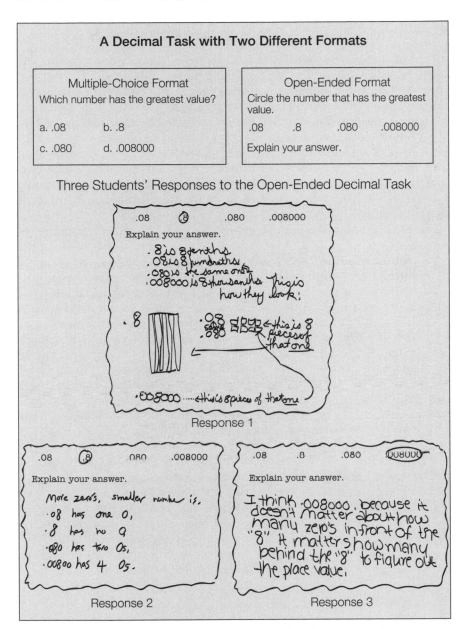

Fig. 18.1. Two formats of a decimal task and three students' responses to the open-ended format

erroneously to explain why .008000 is the number with the greatest value. This student has confused the place values of whole numbers with the place values of decimal numbers.

THE USE OF HOLISTIC SCORING RUBRICS

A holistic scoring procedure is used to score students' responses to the open-ended tasks in the QUASAR Cognitive Assessment Instrument (Lane 1993). In particular, each student's response is assigned a score level ranging from 0 to 4 on the basis of a set of specified criteria. Mathematical communication is one of the three interrelated components in the criteria; the other two components are mathematical knowledge and strategic knowledge. Figure 18.2 shows the general holistic scoring rubric. For each task, the general holistic scoring rubric guided the development of a specific scoring rubric to score students' responses. For students to receive a high-level score, they must clearly communicate their mathematical thinking and reasoning.

Figure 18.3 shows a graph-interpretation task in which students are shown a graph of time and speed and are asked to use the information to write a story about Tony's walk to his grandmother's house. This task is designed to have students communicate their interpretation of the graph by writing stories. Figure 18.3 also shows a set of student responses for each of the five score levels. To receive a score level of 4, a student's story must integrate time and speed for each interval. At the score level of 3, a student's story should display correct and complete interpretation of time and speed except for a minor error, omission, or ambiguity. To receive a score level of 2, a student's story should show some success, albeit limited, in integrating time and speed to interpret the graph. If a student does not integrate time and speed but makes a meaningful reference to the graph, the response would be scored as 1. If a student's story shows no understanding of the graph, the response would receive a score of 0.

When scoring students' responses, the assessor's focus should be on the mathematical nature of the communication rather than on the linguistic nature of the communication. In fact, if a student's explanation for a task is linguistically sound but not mathematically sound, that student would not receive a high score. The low score reflects the fact that the student did not display mathematical understanding of the particular task. Conversely, if a student's explanation is mathematically correct and complete but not linguistically elegant, that student would receive a high score.

CONCLUDING COMMENTS

This paper addresses the role of open-ended tasks and holistic scoring rubrics in assessing students' mathematics communication skills. In the assessment of mathematical proficiency, it is essential that students understand the criteria used for evaluating their performance. Thus, teachers should create instructional activities using open-ended tasks and holistic scoring rubrics that will not only enable students to internalize

Score Level 4	Score Level 3	Score Level 2	Score Level 1	Score Level 0
Mathematical Knowledge	*Mathematical Knowledge*	*Mathematical Knowledge*	*Mathematical Knowledge*	*Mathematical Knowledge*
Shows understanding of the problem's mathematical concepts and principles; uses appropriate mathematical terminology and notations; and executes algorithms completely and correctly.	Shows nearly complete understanding of the problem's mathematical concepts and principles; uses nearly correct mathematical terminology and notations; and executes algorithms completely. Computations are generally correct but may contain minor errors.	Shows understanding of some of the problem's mathematical concepts and principles. Response may contain serious computational errors.	Shows very limited understanding of the problem's mathematical concepts and principles; may misuse or fail to use mathematical terms. Response may contain major computational errors.	Shows no understanding of the problem's mathematical concepts and principles.
Strategic Knowledge	*Strategic Knowledge*	*Strategic Knowledge*	*Strategic Knowledge*	*Strategic Knowledge*
May use relevant outside information of a formal or informal nature; identifies all the important elements of the problem and shows an understanding of the relationships among them; reflects an appropriate and systematic strategy for solving the problem; and gives clear evidence of a solution process. Solution process is complete and systematic.	May use relevant outside information of a formal or informal nature; identifies the most important elements of the problem and shows a general understanding of the relationships among them; and gives clear evidence of a solution process. Solution process is complete, or nearly complete, and systematic.	Identifies some important elements of the problem but shows only limited understanding of the relationships among them. Gives some evidence of a solution process, but the solution process may be incomplete or somewhat unsystematic.	May attempt to use irrelevant outside information; fails to identify important elements or places too much emphasis on unimportant elements; may reflect an inappropriate strategy for solving the problem; and gives incomplete evidence of a solution process. Solution process may be missing, difficult to identify, or completely unsystematic.	May attempt to use irrelevant outside information; fails to indicate which elements of the problem are appropriate; and copies part of the problem but without attempting a solution.
Communication	*Communication*	*Communication*	*Communication*	*Communication*
Provides a complete response with a clear, unambiguous explanation or description; may include an appropriate and complete diagram; communicates effectively to the identified audience; presents strong supporting arguments that are logically sound and complete; and may include examples and counterexamples.	Provides a fairly complete response with reasonably clear explanations or descriptions; may include a nearly complete, appropriate diagram; generally communicates effectively to the identified audience; presents supporting arguments that are logically sound but may contain some minor gaps.	Makes significant progress toward completion of the problem, but the explanation or description may be somewhat ambiguous or unclear; may include a diagram that is flawed or unclear. Communication may be somewhat vague or difficult to interpret, and arguments may be incomplete or may be based on a logically unsound premise.	Has some satisfactory elements but may fail to complete or may omit significant parts of the problem; may include a diagram that incorrectly represents the problem situation, or diagram may be unclear and difficult to interpret. Explanation or description may be missing or difficult to follow.	Communicates ineffectively; may include drawings that completely misrepresent the problem situation. Words do not reflect the problem.

Fig. 18.2. The QCAI General Holistic Scoring Rubric

Graph Interpretation Task

Use the following information and the graph to write a story about Tony's walk.

At noon, Tony started walking to his grandmother's house. He arrived at her house at 3:00. The graph below shows Tony's speed in miles per hour throughout his walk.

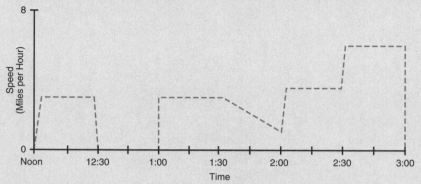

Write a story about Tony's walk. In your story, describe what Tony might have been doing at the different times.

Student Responses

Level 4

This student correctly identified and interpreted the speed using miles per hour and relevant activities for each time interval.

Tony left his house at noon. From noon to 12:30 he was walking about 3½ miles per hour because he was trying to catch ants. At about 12:30 Tony stopped to rest and eat lunch. Then at 1:00 he started walking again. He was walking about 3½ miles per hour until about 1:30 because he was daydreaming. At 1:30 he saw five birds hopping along the sidewalk so he watched them while he was walking which slowed him down a little. Then at 2:00 he finally caught up with his regular pace until 2:30 when he realized the time then he ran the rest of the way.

Level 3

This student correctly interpreted the graph by indicating the speeds or relevant activities for almost all time intervals. This student did not indicate that the speed in the 2:30 to 3:00 time interval was higher than the speed in the 2:00 to 2:30 time interval.

One day a boy was sent to his grandmas house to pick up something he left at 12:00 noon he had know money so can catch the bus. At 12:30 he stoped to rest at 1:00 he started cwalking again at 1:30 - 2:00 he start slowing down until he saw a dog. The dog start chaseing him 2 - 2:30 his owner grabed him bat at 3:00 he was at his grandmas and told her the whole story

Level 2

This student showed some success in integrating the time and speed to interpret the graph.

he stoped somewhere at 12:30 Started back at 1:00 Slowed down at 2:30 at about 2:02 he kept pcking up the pace and he made it at 3:00

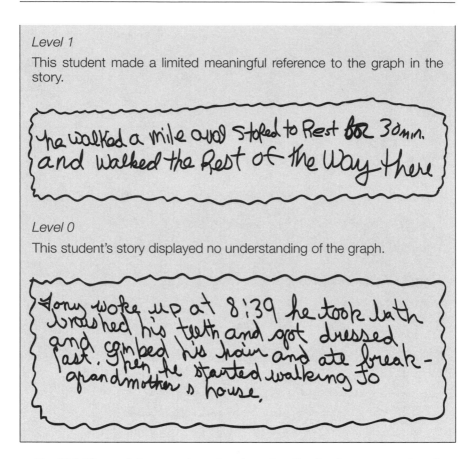

Level 1

This student made a limited meaningful reference to the graph in the story.

> he walked a mile and stoped to Rest for 30 min. and walked the Rest of the way there

Level 0

This student's story displayed no understanding of the graph.

> Tony woke up at 8:39 he took bath brushed his teeth and got dressed and combed his hair and ate break-fast. Then he started walking to grandmother's house,

Fig. 18.3. The graph interpretation task and samples of students' responses to the task

the criteria used for evaluating their performance but also promote increased student proficiency in mathematics. The examples presented here may help mathematics teachers create classroom environments that foster mathematical communication. In particular, teachers can use the general scoring rubric as one suggestion for scoring students' responses. Teachers can use the general scoring rubric to inform the students of the criteria for evaluating their responses and to score the students' responses, as well as use the sample student responses to show the nature of mathematical communication. Teachers can also ask students to identify what is needed to make a low-level response meet the criteria for a high-level response. Doing so will facilitate students' learning of mathematical and strategic knowledge and improve their mathematical communication.

REFERENCES

Lane, Suzanne. "The Conceptual Framework for the Development of a Mathematics

Performance Assessment Instrument." *Educational Measurement, Issues and Practice* 12 (April 1993): 16–23.

Magone, Maria E., Jinfa Cai, Edward A. Silver, and Ning Wang. "Validating the Cognitive Complexity and Content Quality of a Mathematics Performance Assessment." *International Journal of Educational Research* 12 (July 1994): 317–40.

Silver, Edward A. *Quantitative Understanding: Amplifying Student Achievement and Reasoning.* Pittsburgh, Pa.: Learning Research and Development Center, 1993.

19

Mathematical Communication in Students' Responses to a Performance-Assessment Task

Dominic Peressini
Judy Bassett

THE ability to communicate mathematics is essential as students strive to develop their mathematical power. Through communication the student and the teacher are able to exchange mathematical knowledge. This communication, in turn, shapes instructional and assessment practices. Performance assessment in mathematics promotes this exchange of information. This paper examines a number of responses to a performance-assessment task that engages students in the use *and* communication of mathematics.

Linkages between mathematical reasoning and communication are central to many important initiatives in the reform of mathematics education in the United States. Indeed, Hiebert's (1992) insightful analysis of reform shows that the processes of reflection and communication have provided the framework and much of the direction for the restructuring of mathematics education. More particularly, since communication embodies the social aspect of learning mathematics, Hiebert (p. 446) argues that

> communication can promote and guide reflection, and reflection can enrich what is shared through communication.

This emphasis on communication is reflected in NCTM's (National Council of Teachers of Mathematics) *Curriculum and Evaluation Standards*

The Wisconsin Performance Assessment Development Project is developing performance-assessment tasks for the state of Wisconsin as part of the Wisconsin Student Assessment System. Research and development is conducted at the Wisconsin Center for Education Research under the direction of Norman Webb. The authors recognize Wesley Martin for his assistance with this project and his insightful suggestions concerning this paper.

for School Mathematics (1989), which establishes communication as one of the Standards for school mathematics at all levels. The *Professional Standards for School Mathematics* (NCTM 1991) further reflects the importance of communication in the teaching of mathematics. In particular, teachers are encouraged to choose and develop tasks that foster their students' ability to communicate mathematically.

Performance-Assessment Tasks

One such task that engages students in problem solving, reasoning, and communication is the mathematics performance-assessment task. As mathematics reform has proceeded, the performance-assessment task has become one of the more promising tools for assessing students' understanding of mathematics (Romberg 1993). Performance-assessment tasks allow students to communicate their mathematical knowledge in an authentic manner that is meaningful to their life.

Stenmark's (1991, p. 13) description of performance assessment in mathematics bears directly on the use of performance-assessment tasks to assess students' understanding:

> A performance assessment in mathematics involves presenting students with a mathematical task, project, or investigation, then observing, interviewing, and looking at their products to assess what they actually know and can do.

Moreover, Stenmark has suggested that quality tasks should be essential, authentic, rich, engaging, active, feasible, equitable, and open. These desirable features work together to elicit responses that communicate students' mathematical knowledge.

When performance tasks are used for assessment, it is imperative that students communicate their reasoning when they respond to the tasks. Such communication not only enriches students' learning but also assists the teacher in making an accurate assessment of students' thinking and, as a result, allows the teacher to make appropriate instructional decisions for individual students.

Students' Responses to a Performance-Assessment Task

The remainder of this article examines responses to one performance-assessment task in which students are required to communicate their mathematical reasoning. The students are fourth graders from Wisconsin; they range in age from seven to nine years. The examples of students' responses are taken from a task developed as part of the Wisconsin Performance Assessment Development Project. This task—which meets Stenmark's criteria for a quality performance-assessment task—is one of many tasks of this kind that have been written, developed, pilot-tested, and scored by mathematics teachers throughout the state of Wisconsin.

The Task

The Race (fig. 19.1) is a task that is set in a context familiar to elementary school children—running a race on a school playground. The task uses their interest in the notion of fairness to engage them and motivate them to solve the problem it poses. Students are given two different race plans to consider for fairness, and they are asked to support their findings through written communication. The Race is open-ended in that the task can be solved in multiple ways, the task has multiple solutions, and the final requirement of the task asks students to design their own race plan and describe whether it is fair or not.

Analysis of Students' Communications

The first step in analyzing students' responses was to score each student's solution using a holistic scoring rubric. This rubric served as a framework for consistently evaluating and scoring students' solutions. Most students were able to make a correct decision regarding the fairness of this race. However, by analyzing a student's communication, it is possible to shed light on many aspects of the student's thinking and solution processes beyond the correct decision.

Mathematical reasoning

Josh, like most students, approached this task by using a ruler (see fig. 19.2). Although using a ruler is an appropriate method for solving this task, the method leaves unanswered questions about the student's basic understanding of the length of a diagonal versus the length of a side of a square. Josh's approach makes clear neither whether he has a grasp of the relationship between the diagonal and the side of a square nor whether he did not feel confident enough in that knowledge to use it to support his decision. He also stated that the racers ran the literal measurement of the diagram (i.e., 1.5 and 2 inches). This mistake may have been an oversight, or Josh may not have realized the connection between the diagram and the real world it was modeling. If he understood that the diagram represents a longer distance (a scale drawing), he did not communicate that understanding.

Brooke, conversely, stated that the race was unfair because the runners did not run an equal distance (see fig. 19.2). Brooke correctly indicated that Dolores had to run farther and hence responded appropriately to the requirements of the task. She did not, however, furnish evidence about how she reached that decision. Her response does not let the teacher know whether she used geometric knowledge and reasoning, measured with a ruler, estimated on the basis of sight, or used some other method. It would be natural to conclude that Brooke has failed to communicate enough information, but we must be careful when we interpret each response. Another interpretation—and perhaps a more valid one—is that this task, as it is posed, does not elicit enough information from Brooke

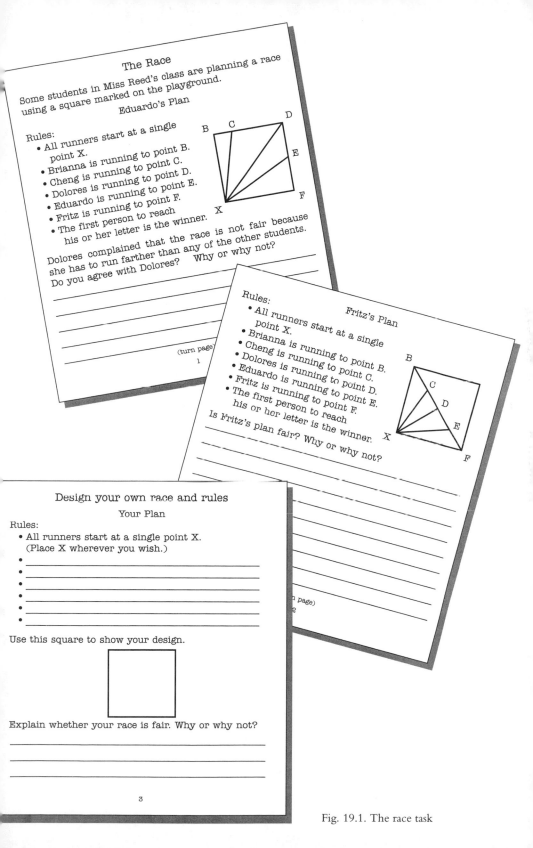

The Race

Some students in Miss Reed's class are planning a race using a square marked on the playground.

Eduardo's Plan

Rules:
- All runners start at a single point X.
- Brianna is running to point B.
- Cheng is running to point C.
- Dolores is running to point D.
- Eduardo is running to point E.
- Fritz is running to point F.
- The first person to reach his or her letter is the winner.

Dolores complained that the race is not fair because she has to run farther than any of the other students. Do you agree with Dolores? Why or why not?

_____
_____
_____
_____
_____

(turn page)

1

Fritz's Plan

Rules:
- All runners start at a single point X.
- Brianna is running to point B.
- Cheng is running to point C.
- Dolores is running to point D.
- Eduardo is running to point E.
- Fritz is running to point F.
- The first person to reach his or her letter is the winner.

Is Fritz's plan fair? Why or why not?

_____
_____
_____
_____

(turn page)
2

Design your own race and rules

Your Plan

Rules:
- All runners start at a single point X. (Place X wherever you wish.)
- _____
- _____
- _____
- _____
- _____

Use this square to show your design.

Explain whether your race is fair. Why or why not?

_____
_____
_____
_____

3

Fig. 19.1. The race task

to enable a teacher to make an informed decision about Brooke's mathematical processes. In this sense it is the task that is inadequate, not the student's response.

Dolores complained that the race is not fair because she has to run farther than any of the other students.
Do you agree with Dolores? Why or why not?

Yes, I agree with Doloras. Brianna, Cheng, Edwardo, and Fritz run 1½ inches Dolores runs 2 inches.

Josh's Response

I think that Dolores is right because she has to run Further then any other Student in The race. That is Not Fair every one should Race exwel the same.

Brooke's Response

Fig. 19.2. Josh's and Brooke's responses, page 1. Josh and Brooke have "correct answers"; their communications, however, leave questions regarding their reasoning.

Teacher-student communication

The analysis of students' communications requires the analysis of the teacher's communication as well. Seth recognized that the distances are unequal, but he based his decision on a different notion of "fairness" than the other students did (see fig. 19.3). Rather than focus on the geometric construction of the racecourse, Seth focused on the individual racers' strengths and weaknesses. He chose to direct his attention to the social aspects of the race, and he decided to make the race "fair" by ensuring that all contestants have an equal chance of winning. His response was unexpected and caused the writer of the task to critically evaluate her own communication skills. The writer of the task assumed that the student would give a mathematical response focusing on the geometric construction of the racecourse. However, that assumption was not communicated successfully to Seth. In retrospect, a review of the task reveals that nothing in the task would disallow Seth's response. In fact, Seth should be commended for his insightfulness and creativity in responding to the task.

Open-ended assessment tasks, which allow students' communications, often lead to unexpected responses. In dealing with these unexpected responses, we must ask the question, Is this unexpected (and perhaps unwanted) response due to the student or the task? In this regard, students' communications can reveal many problems with the construction of an assessment instrument that have been difficult to identify in traditional mathematics tests.

Dolores complained that the race is not fair because she has to run farther than any of the other students.
Do you agree with Dolores? Why or why not?

I don't think it's fair because run farther ther. the rest of the class because she might be a slow runner nobody knows. I think the rase should be changed.

Seth's Response

Fig. 19.3. Seth's response, page 1. Seth concludes that the race is unfair, but his communication reveals that he did not use mathematical reasoning to reach that conclusion. Is this his problem or a deficiency in the task?

Mathematical misconceptions

Analyzing students' communications not only allows the teacher to identify the mathematical development of students but also alerts the teacher to the mathematical misconceptions of students. Carlee apparently thought that as long as all students begin at the same point, the race is fair (see fig. 19.4). She did not reveal enough information on page 1 of her response to permit the teacher to be certain that Carlee was operating under a mathematical misconception, but she did furnish evidence of her problem: focusing solely on the beginning of the race and not taking into account the actual distances each person travels. However, as the teacher studied page 2 of Carlee's response, more evidence was provided that suggests Carlee does have some misunderstanding about the relationships among the lines that can be constructed in a square. On pages 1 and 2, Carlee wrote that either beginning at the same point or ending on the same line made the distances equal. Carlee's response to the performance-assessment task shows that she either confused some basic geometric knowledge or perhaps did not understand the context of the assessment situation. This misunderstanding could derive from social or cultural experiences. For example, Carlee may not have participated in many races because she has been socialized according to cultural norms in which women do not compete in these types of activities. As a result, Carlee may not understand the aspects of a race and the fact that unequal distances give some students an advantage. The teacher must once again be open to a variety of plausible interpretations of each student's response.

Dolores complained that the race is not fair because she has to run farther than any of the other students.
Do you agree with Dolores? Why or why not?

"No" because all kids start at the X, so she would start with the other kids.

Carlee's Page 1

Is fritz's plan fair? Whayor why not?

Yes because they all end up at the same line, so why woudn't it be fair.

Carlee's Page 2

Fig. 19.4. Carlee's responses. Carlee *appears* to have a misconception. Her communication is not sufficiently complete to analyze her reasoning.

Geometric reasoning

In contrast, Katie communicated sufficient geometric information to demonstrate that she can solve the first part of the task (see fig. 19.5).

She used mathematical language to state that a quarter of a circle—rather than a square—would make the race fair. An interview with her revealed that she did indeed have the necessary understanding to reach and support her valid conclusion. However, her written communication skills could be refined in order to maximize her power to express mathematics in prose.

Dolores complained that the race is not fair because she has to run farther than any of the other students.
Do you agree with Dolores? Why or why not?

Yes. The race participants should not be using a square, they should be using a quarter of a circle.
(D) to ensure that everyone has to run the same distance. Dolores has to run to the far corner of the square

---

Fig. 19.5. Katie's response, page 1. Katie uses mathematical terms to communicate her reasoning. Improvement is still needed.

Student Reentry and Teacher Reassessment

The second part of the performance-assessment task gives both students and teachers an opportunity to reconsider the student's initial response. (Page 2 of the task is presented in fig. 19.1.) When Josh and Brooke completed the second part of the task, they showed the teacher that they applied the same reasoning to each of the first two parts of the performance-assessment task (see fig. 19.6). Josh did not give the teacher any new information by completing the second part of the task. However, although Brooke used the same strategy in both parts of the task, some new information was elicited in the second part. Brooke evidently did not recognize that Fritz has to run as far as Brianna and that the other runners are not running equal distances.

Carlee's response (fig 19.4) to the second part of the task communicates additional information that confirms that she has misunderstood either some mathematical concepts or the context of the task. These misunderstandings can therefore be addressed. Reentering the task in the second part makes it possible for a teacher to prompt students to reconsider their original responses, and—of equal significance—reentry makes it possible for the teacher to reassess students' responses from the first part of the task.

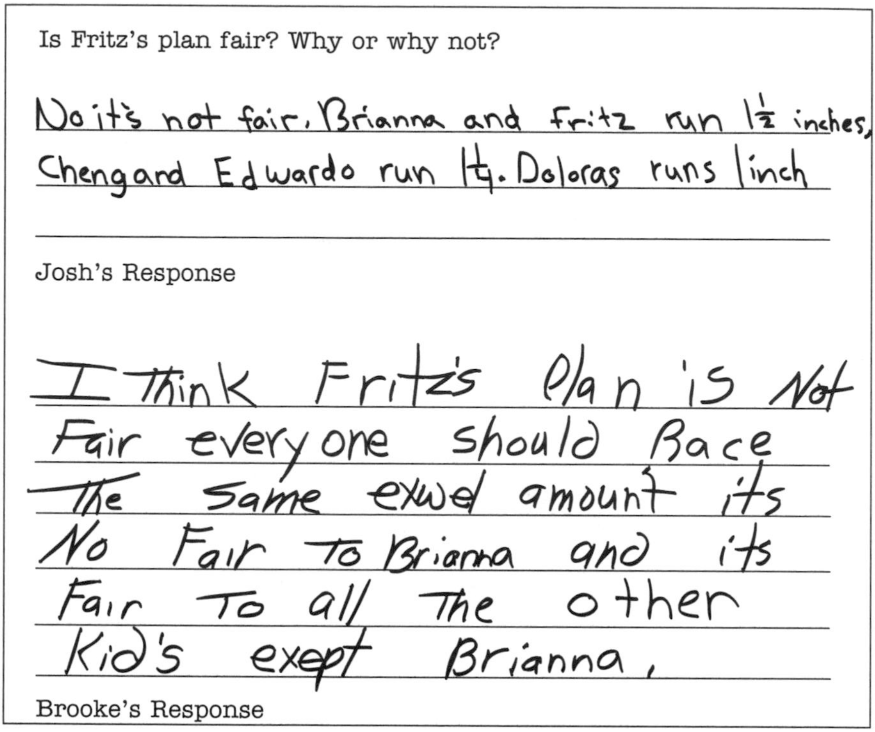

Is Fritz's plan fair? Why or why not?

No it's not fair, Brianna and Fritz run 1½ inches, Cheng and Edwardo run 1¼. Doloras runs 1 inch

Josh's Response

I Think Fritz's Plan is Not Fair every one should Race The Same exwd amount its No Fair To Brianna and its Fair To all The other Kid's exept Brianna,

Brooke's Response

Fig. 19.6. Josh's and Brooke's responses, page 2. Josh and Brooke use the same justification for the second part of the task that they used on the first part. What new information is revealed about these students?

The Sophistication of Mathematical Communication

The third part of the performance-assessment task is entirely open ended, inasmuch as it asks the student to define both the starting and ending points of the race (page 3 of the task is presented in fig. 19.1). This part is open both in the sense that students are free to make their own designs and rules and in the sense that the students are not required to make them "fair." What they must do is clearly communicate their reasoning in developing the race and be able to explain their judgment of its fairness. In conjunction with the previous two parts of the assessment task, this portion of the task enables students to show their teacher different levels of sophistication in mathematical communication.

Danna solved the problem by using a ruler (see fig. 19.7). She combined the literal measurement of one inch with language representing how far the students have to run. She clearly recognized that the line from the center of the square that is not perpendicular to the sides of the square would be longer if it were extended to the sides of the square. She used the term *angle* to describe Fritz's path. However, her communication

Use this square to show your design:

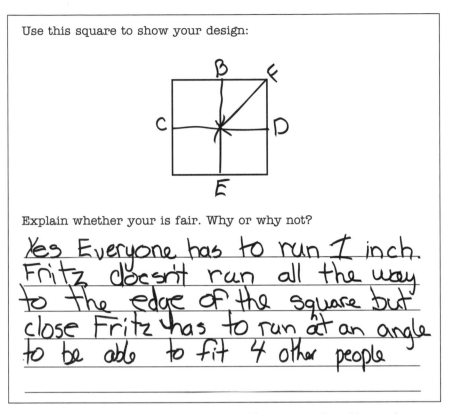

Explain whether your is fair. Why or why not?

Yes Everyone has to run 1 inch.
Fritz doesn't run all the way
to the edge of the square but
close Fritz has to run at an angle
to be able to fit 4 other people

Fig. 19.7. Danna's response, page 3. Danna provides evidence of equal lines and uses some mathematical language. We don't know if she simply measured or if she also understood and used geometric concepts to design her plan.

does not reveal whether she also realized that the remaining racers were running at an angle or if she recognized the relationships between the lines representing the racers' distances. It could be that Danna's perspective was confounding her interpretation of the race. Nonetheless, without further communication, her teacher is unable to judge whether Danna is able only to measure the distances accurately or whether she also understands the mathematical underpinnings of her plan.

Westin and Jerrod had similar plans (see fig. 19.8). In assessing these students' work, their teacher discovers that they both said the same thing. The fact that Westin connected the points to show a quarter circle or that Jerrod wrote more is not compelling evidence of a deep mathematical understanding. It is not evident that either Westin or Jerrod reflected mathematically about *why* his race was fair. Each has merely stated that the distances are equal. With practice designed to improve mathematical communication, it is entirely possible that many students could give more conclusive evidence of sophistication in the application of geometric principles.

Use this square to show your design:

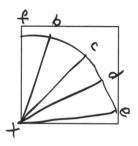

Explain whether your design is fair. Why or why not?

its fair because all runers havean
eoel amount to run

Westin's response

Use this square to show your design:

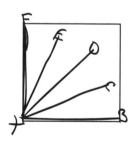

Explain whether your design is fair. Why or why not?

I think it's fair because every student
runs the same distance, and no one
student's path crosses another
students path, So the won't get in
each others way.

Jerrod's response

Fig. 19.8. Westin's and Jerrod's responses, page 3. As you view Westin's and Jerrod's responses, beware of the appearance trap. Only the communication can be assessed.

Concluding Comments

Analyzing communication in mathematics can be beneficial to both students and teachers. We must, however, be cautious about the role of communication in teaching and assessment. Without communication in mathematics, we have little evidence of students' understanding and application of mathematical processes. With communication, we must be careful not to assume that students understand a mathematical concept or process when they have not clearly expressed themselves on the matter. We need to exercise caution, however, in concluding that a student lacks understanding of a concept if he or she does not communicate that understanding in an assessment situation. We need to keep in mind that through mathematical communication we are assessing what students know and are able to do—rather than what they do not know or are not able to do.

In analyzing students' communications, we need to question whether the weaknesses we note indicate that a given student needs further instruction and development of mathematical knowledge, that the student needs additional help in communicating the mathematical knowledge that he or she already possesses, or that the assessment task itself is inadequate for what we are trying to measure. By asking students to explain their solutions—both orally and in writing—the teacher gains an opportunity to distinguish among mathematical misconceptions, differences in social and cultural experiences, and communication problems. In any event, the teacher has learned something about the students that is not generally revealed through traditional assessment techniques that offer little opportunity for student communication.

By means of the performance-assessment process, the teacher can learn that her or his students need to develop the communication of their mathematical knowledge. The teacher could begin the process of development by showing students how to ask questions that will clarify their thinking and communicate their ideas more clearly. The students should practice asking questions of their peers in small groups to elicit clearer responses. If students are to value understanding and communication in mathematics, they need practice in the art of questioning. In addition, students can be shown actual samples of effective written communication, and the teacher can organize class discussions to analyze the characteristics that make each sample effective. Finally, students could be asked to revise each sample so as to improve it. Just as students have edited and revised their writing in language arts classes for years, they must also edit and revise their communications in mathematics classes.

Perhaps the greatest danger in analyzing students' responses is the tendency to accept one interpretation of a student's work. The previous examples of students' work demonstrate that a number of valid interpretations can be made of the same response. Both as teachers and researchers, we must be careful to guard against our personal biases, and we must acknowledge that there may be many interpretations of a student's work.

As Hiebert (1992) suggests, communication can generate and direct reflection. He defines reflection as "the conscious consideration of one's experiences" (p. 440). This consideration includes mulling, distilling, contemplating, and monitoring. This notion of reflection should not be confined to the student; it should also include the teacher. Communication allows the teacher to reflect on the individual student's mathematical reasoning and to make informed instructional decisions that are specific to that student's needs. In this manner, communication becomes an essential part of the student's learning of mathematics.

In realizing that performance-assessment tasks provide opportunities for students to communicate mathematically, we are assisting in the effort to develop mathematical power for all students. The *Assessment Standards for School Mathematics: Working Draft* (NCTM 1993, p. 5) states that

> for all students to achieve mathematical power, they need to become mathematical problem solvers, to value mathematics, to *reason and communicate* mathematically, and to be confident in using mathematics to make sense of real-world problem situations. The mathematics curriculum, the instructional methods, and the strategies used to *assess* student performance must be congruent with this notion of mathematical power [emphasis added].

As this congruence continues to develop, students will learn how to develop and use their mathematical power.

REFERENCES

Hiebert, James. "Reflection and Communication: Cognitive Considerations in School Mathematics Reform." In *International Journal of Educational Research,* vol. 17, edited by Walter G. Secada, pp. 439–56. Oxford: Pergamon Press, 1992.

National Council of Teachers of Mathematics. *Assessment Standards for School Mathematics: Working Draft.* Reston, Va.: National Council of Teachers of Mathematics, 1993.

―――. *Curriculum and Evaluation Standards for School Mathematics.* Reston, Va.: National Council of Teachers of Mathematics, 1989.

―――. *Professional Standards for Teaching Mathematics.* Reston, Va.: National Council of Teachers of Mathematics, 1991.

Romberg, Thomas A. "How One Comes to Know: Models and Theories of the Learning of Mathematics." In *Investigations into Assessment in Mathematics Education,* edited by Morgan Niss, pp. 97–111. Dordrecht, Netherlands: Kluwer Academic Publishers, 1993.

Stenmark, Jean Kerr. *Mathematics Assessment: Myths, Models, Good Questions, and Practical Suggestions.* Reston, Va.: National Council of Teachers of Mathematics, 1991.

20

Communication Processes in Mathematical Explorations and Investigations

Carole Greenes

Linda Schulman

In the past decade, leaders in the core curricular areas have been actively promoting the development of "habits of mind" in their disciplines. The *Curriculum and Evaluation Standards for School Mathematics* (NCTM 1989) has identified reasoning, problem solving, and communicating as processes that are important to the learning of mathematics and to the solving of mathematical problems. Scientists and science educators have indicated that the processes associated with the scientific method (e.g., observing, organizing, analyzing, classifying, communicating) are at the heart of learning and doing science. History in the schools is now viewed as an investigative discipline rather than a corpus of facts. Similarly, literacy and English are being considered investigative disciplines in which writing, reading, speaking, and listening are presented as interrelated processes. These "habits of mind" have also been identified as essential to the successful performance of workers in all fields and occupations (Secretary's Commission on Achieving Necessary Skills 1991).

A common need in all disciplines and in the workplace is the ability to communicate. All core subjects and all vocations require that people be able to—

- express ideas by speaking, writing, demonstrating, and depicting them visually in different types of displays;
- understand, interpret, and evaluate ideas that are presented orally, in writing, or in visual forms;
- construct, interpret, and link various representations of ideas and relationships;

- make observations and conjectures, formulate questions, and gather and evaluate information;
- produce and present persuasive arguments.

PROMOTING COMMUNICATION: EXPLORATIONS AND INVESTIGATIONS

Because communication has been shown to be not only a way of clarifying students' thinking and understanding but also a way of revealing their thinking, their reasoning, and what they know and do not know, vehicles are needed for promoting communication. Two such vehicles that we have found to be successful are short-term explorations and long-term project types of investigations.

Short-Term Explorations

Short-term explorations are designed to be used in the classroom and may be completed by students working alone or with a partner. The explorations require students to describe observations, justify solutions, or document the steps they followed in their thinking to resolve a problem (Greenes, Schulman, and Spungin 1992). For these explorations, all the necessary data are included in the problem statement, except those that the student may generate or collect in the classroom. Students record their observations (in, e.g., drawings, text, tables, graphs), justifications, and documentations in their mathematics journals, then discuss their work and receive feedback on it. Examples of each of these types of explorations follow, along with students' responses.

Describe observations

The students in a grades 1–2 class explored this problem by working in pairs:

Draw pictures of the people who made these footprints. Tell how the people are the same. Tell how the people are different.

Each pair of students received a copy of the footprints picture and another sheet of paper with two columns. At the bottom of one column was the small footprint; at the bottom of the other column was the large footprint. In the columns above the prints the children drew pictures of the people they thought made the footprints. After completing their drawings, students described, compared, and contrasted the people. A snapshot of their discussions follows:

Dana: The small feet are children. The big feet are parents.

Jennifer: I think the little feet are mothers and the big feet are fathers. My father has real big feet. He can't buy shoes in a store.

Bobby: My mother has big feet, too. They are bigger than Dad's feet.

Dana: My mother doesn't have big feet, but she is tall.

Lisa: My father is not tall for a man.

Jennifer: Well, I think for sure that the small feet have to be children's.

Lisa: So the big feet belong to someone who is older.

From this discussion, it is clear that the students were able to identify a number of variables—including age, height, and gender—that might contribute to the differences in foot size. As the students described, compared, and contrasted their drawings, the teacher recorded the variables on the chalkboard, then focused the discussion on ways to check the relationship between foot size and height and between foot size and gender.

Justify solutions

Multiple answers to this question are possible, as students in a grade 5 class noted after much discussion:

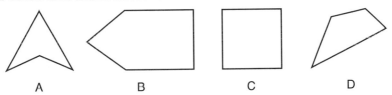

Which shape does not belong with the others? Why? Could there be a different answer? Explain.

All the initial responses were that Shape B did not belong because it was the only figure that did not have four sides. When prompted to consider other solutions, the students gave the following answers and explanations:

Elliot: D doesn't belong. It doesn't have any two sides that are the same length.

Gina: C doesn't belong because it is the only square. And it's the only one that has all four sides the same length. All four angles are right angles. None of the others have that.

Raiza: I think D doesn't belong for a different reason. It doesn't have any angles that are the same like the others do. All of its angles are different.

At this point, no student had decided to exclude Shape A. With prompting from the teacher, Deirdre responded, "Well, I'd say A is different from the other shapes. I don't know the math reason. But you could say that it's pushed in. It looks different from the other shapes." After

Deirdre's comment, the students began to look for new reasons for excluding each shape. Gina added, "You could say C doesn't belong because it is the only shape with four lines of symmetry." Building on the concept of symmetry, the students examined each of the other shapes and noted that Shapes A and B each have one line of symmetry and Shape D has zero lines of symmetry.

From the students' responses, the teacher gained insight into what they knew about quadrilaterals and where they needed assistance. Although the students were able to distinguish quadrilaterals from non-four-sided figures, they did not have the vocabulary to describe angles of the same size and sides of the same size, nor did they know the terms *convex* ("pushed out") and *concave* ("pushed in"). The students did, however, have the concept and language of symmetry.

Document steps in thinking

Ninth-grade students approached this problem in two different ways:

> Amy has the same number of sisters as brothers. Her brother Jason has twice as many sisters as brothers. How many children are in the family? Describe the thinking steps you followed to solve the problem. (Greenes et al. 1980)

The majority of students reasoned through the problem by generating guesses, checking guesses with the problem conditions, and modifying guesses and checking again until all the conditions were met. However, it was noted by the teacher that many of these students attempted an algebraic approach first, became confused in the counting of "sisters to Jason" and "sisters to Amy," and then resorted to the guess-and-check strategy. A second, much smaller group of students used an algebraic approach successfully.

Jarret's guess-and-check thinking steps

Step 1: First I tried to guess. I said Amy has the same number of sisters and brothers. So Amy could have 1 sister and 1 brother.

Step 2: Then I checked with Jason. It doesn't work. Jason has 2 sisters and no brothers, and Jason has to have twice as many sisters as brothers. It's tricky because you have to count Jason as Amy's brother but not his own brother.

Step 3: I got stuck so I made a table of boys and girls. I wrote Amy and Jason in the table. (Girls: Amy   Boys: Jason)

Step 4: Amy has to have the same number of sisters and brothers, so I put in 1 more sister. That works for Amy but not Jason. (Girls: Amy, sister   Boys: Jason)

Step 5: Then Jason has to have a brother, so I put in 1 more boy. That works for Jason but not for Amy. (Girls: Amy, sister   Boys: Jason, brother)

Step 6: Amy has to have the same number of sisters as brothers, so I added 1 sister. It works for Amy. She has 2 sisters and 2 brothers.

It still doesn't work for Jason. He has 3 sisters and 1 brother. (Girls: Amy, sister, sister   Boys: Jason, brother)

Step 7: So I tried to work on Jason. He has to have twice as many sisters. So I tried 4 sisters and 2 brothers. It works. Amy has 3 sisters and 3 brothers. Jason is 1 of the brothers. Jason has 4 sisters with Amy and 2 brothers. There are 7 children in the family.

Emma and Janice's algebraic step-by-step solution

Step 1: Let x = number of Amy's sisters. That means the number of girls is $x + 1$. The 1 is for Amy.

Step 2: Let y = number of Jason's brothers. The number of boys is $y + 1$. The 1 is for Jason.

Step 3: From Amy's condition, the number of boys has to equal x.

Step 4: From Jason's condition, the number of girls has to equal $2y$.

Step 5: With both conditions, the number of girls equals $x + 1 = 2y$. The number of boys equals $x = y + 1$.

Step 6: We solved the equations $x + 1 = 2y$ and $x = y + 1$:

$$x = y + 1$$
$$(y + 1) + 1 = 2y$$
$$y + 2 = 2$$
$$y = 2$$
$$x = 3$$

Amy has 3 sisters, so there are 4 girls. Jason has 2 brothers, so there are 3 boys. It checks. There are 7 children.

Long-Term Investigations

Long-term investigations require students to engage in the investigative process involving (1) the formulation of questions based on prior knowledge or observations, (2) the collection of data and other relevant information, (3) the organization of data and information for inspection and analysis, (4) the analysis of data, and (5) the presentation of results and conclusions. Each of these processes can be visited any number of times during the conduct of an investigation. Each of these five processes involves communication through speaking, writing, visual presentations, or electronic technology.

Popcorn investigations

Mrs. Perez teaches fifth-grade mathematics in a small urban school. She initiated a discussion with her students about popcorn. Who had eaten popcorn? Who liked popcorn? What are the different types of popcorn? How can mathematics help us learn more about popcorn? As students offered ideas in response to the last question, Mrs. Perez encouraged students to elaborate on one another's ideas with questions such as Does this idea suggest anything else? and Is there more you would like to know about this? Throughout the brainstorming session, Mrs. Perez recorded the students' ideas on mural paper in a curriculum web (see fig. 20.1).

Mrs. Perez realized that there was not sufficient time to have the students pursue all the questions on the web. Some of the questions were richer mathematically than others. Mrs. Perez decided that the first investigation would involve students in the construction of experiments dealing with the popping of popcorn. Before presenting the guidelines for the investigation, Mrs. Perez arranged for the microwave oven to be moved from the teacher's room to her classroom for two weeks. She also brought in a toaster oven. Four students brought in air poppers from home for the two weeks. A hot plate was loaned by the science teacher.

At the beginning of the first week, Mrs. Perez described the operation of the three cooking sources and gave the students bags of popcorn kernels of various prices and two special bags of popcorn. One special bag contained kernels that had been soaked overnight the previous night. The kernels in the other special bag had been heated at a temperature of 200°F (93°C) in the toaster oven for one hour before being placed in the bag on the first day of the investigation. All kernels were of the same brand. Students were assigned the task of designing and conducting one or two experiments using these materials, as well as any other materials readily available in the classroom. The design of the experiments, the observations and

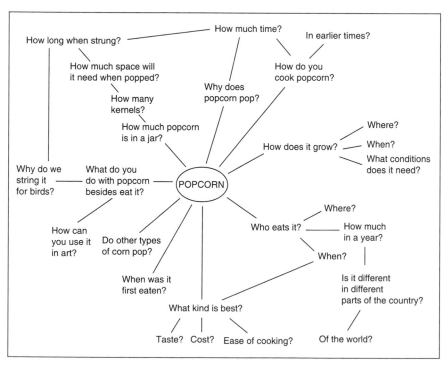

Fig. 20.1. Popcorn curriculum web

hypotheses, and the data-collection techniques as well as the data, the results of the experiment, and the conclusions were to be recorded by the students in their mathematics journals.

The students discussed the possible experiments for about fifteen minutes. During this time, conjectures were made, plans were developed, and needs for materials were identified. The students then began to work in their exploration groups of four to five students each. One group popped thirty kernels from each bag in the air poppers and compared the number of "flakes" (i.e., popped kernels) and "duds" that resulted. The students used graphing software to construct circle graphs to present their findings. The same group of students then compared the cost of the popcorn to the number of duds; conducted the same experiment with popcorn of different prices; and concluded that more expensive popcorn results in fewer duds.

A second group of students compared the numbers of duds resulting from popping the unheated and unsoaked kernels with the numbers resulting from popping the soaked kernels. A third group compared the untreated kernels with the heated kernels. The heated kernels made a great deal of noise when they were cooking on the hot plate. The students were surprised when they removed the lid to find burned kernels.

A fourth group of students conducted a taste-preference survey, comparing the same brand of popcorn prepared by the three different cooking methods. Still another group compared the cooking times for the three different methods. After all the experiments were completed, the students described what they had done, talked about what they had learned, and discussed how to control variables in an experiment.

For the next set of investigations, Mrs. Perez directed students' attention to the curriculum web they had developed. She asked each group to focus on a question that had not yet been examined and to plan an investigation to answer the question selected.

Supermarket investigations

The purpose of the supermarket investigations was to have students in grades 6 and 7 develop strategies for collecting, organizing, displaying, and analyzing data; identify factors that could affect the solution to the problem and modify experiments to take into account those factors; develop interview and group problem-solving skills; and practice making measurements and computing averages. Because most of the data had to be obtained from a supermarket, arrangements were made for field trips led by parents and high school students after school hours. Supermarkets were contacted well in advance, and the investigations and requirements were described in detail.

The following question was posed:

> During a peak hour, which is the faster check-out lane, the express lane or the regular lane?

To begin this investigation, the students called a supermarket manager to get permission to conduct their exploration and to obtain information about peak hours and times when both the express and the regular lanes were open. The manager explained that peak hours, the times when the numbers of customers are greatest, are generally between 4:00 P.M. and 7:00 P.M. daily, including Saturday.

Before leaving for the market, the students discussed the number of hours they would need to collect data, the number of customers they would time, when the timing would start and stop, who would do the timing, and who would be responsible for recording the data. The students decided they would collect data over a period of three days—Tuesday, Friday, and Saturday—and would try to time twenty people in each type of lane each day. One student suggested just counting the number of people who go through each lane in an hour, then dividing to find the average amount of time for each person. The other members of the group disagreed, since they believed that people often leave lines or change lines and that for those reasons, it would be better to follow individual customers. The latter plan was adopted.

After collecting and compiling their data, the students engaged in a discussion that lasted three days. During that time, they identified many variables that could contribute to variations in the speed of checkout. Their list of variables included the following:

- A bagger in the lane in addition to the clerk.
- Composition of the grocery bag. Some bags hold more, like paper. Plastic bags hold less. Some people want paper in plastic and that takes time.
- Speed of the bagger; speed of the clerk.
- Size of items affects the number of bags, and it takes time to open bags.
- Weights of items because baggers try to make bags easy to carry, so they may take time sorting items, or may use more bags.
- Some items, like fresh fruits and vegetables, have to be weighed at check-out.
- Some items have bar codes for scanning and sometimes the clerk can't find the bar code right away. Some items don't have bar codes so the clerk has to enter everything.
- People pay in different ways: credit card, cash, checks. Checks take the longest.
- Some people bring more items into the express lane than they are supposed to. One clerk was upset with a customer who had too many items. The customer said, "It's three cans for $2.29. That's one price. So it should count as one item."
- Customers are always changing lanes. It takes time to move from one lane to another.

After completing and discussing the list, groups of students devised new investigations to determine the influence of some of these variables on the check-out speed. Their investigations were used as extended learning experiences after school. For example, one group of students decided to examine the relationship between the form of payment and the speed of checkout. The students focused on customers in regular check-out lanes during peak hours and timed their financial transactions from the moment the clerk announced the total of the purchases to the customer to the time the clerk handed the receipt to the customer after payment was made. The transaction times were recorded to the nearest second for three groups of customers: forty cash-paying customers, forty customers using bank or credit cards, and forty customers paying by check. The students first displayed their data using three line plots, one for each form of payment. Subsequently, for each group of data, the students found the median of all the data, the median of the first half of the data (the first quartile), and the median of the second half of the data (the third quartile) and used these medians to construct a box plot. From the three box plots, the students were able to compare the distributions of the data and draw conclusions about the form of payment that resulted in the fastest checkout.

Another group of students investigated the relationship between the use of baggers and the check-out time. During the same thirty-minute period in peak shopping hours, the students counted the number of customers that checked out of each of two regular lanes, one with a bagger in addition to the clerk and one with only a clerk. The students repeated the experiment four times. They then computed the mean number of customers in each type of lane for the five thirty-minute periods and compared the means by expressing the difference in means as a percent increase in the number of customers that could be served in a lane with a bagger over the number that could be served in a lane without one. The students wrote to the manager of the supermarket with their recommendation for baggers in addition to clerks and supported their recommendation with the data from their investigation.

The Role of the Teacher

As stated in the *Professional Standards for Teaching Mathematics* (NCTM 1991, p. 57), a learning environment that allows for serious mathematical thinking requires "a genuine respect for others' ideas, a valuing of reason and sense-making, pacing and timing that allow students to puzzle and to think, and the forging of a social and intellectual community." Such a learning environment requires of the teacher very different skills and decisions than those of a traditional classroom. Teachers must facilitate the exploration-and-investigation process through their own behaviors and the ways they organize their classroom.

In order to stimulate interest in explorations and investigations, teachers must be able to identify engaging questions and stimulate students to ask such questions. Modeling such inquiry types of behavior is essential for students to be able to adopt this type of behavior and for it to become a habit. Once this type of behavior is ingrained, students will make comments and ask questions such as these: I wonder why this happens! Will this always be true? Are other answers possible?

In explorations and investigations, classroom authority is shared by teachers and students. Teachers often report to us that the most difficult part of establishing an investigation-based learning community is knowing when to intervene. Tension may develop between the desire to allow students the freedom to pursue their own interests, make their own decisions, reach an impasse, and make mistakes and the desire to help students gain the most they can from the learning experience without "suffering." Over time, teachers gain confidence in their ability to know when to step back and when to intercede. Consider the words of a third-grade teacher who has been experimenting with explorations and investigations for about one year:

> At first I didn't know when to give the children help and when not to. It was hard to just watch the children do something that I knew wouldn't lead anywhere. I always wanted to rush in and help them. I began to train myself to wait longer before plunging in. Sometimes I would walk away from a group that was stuck. Often, by the time I got back to the group, the students had found their own way to continue. When I saw the satisfaction the children felt when that happened, I knew that I was doing the right thing. Now, I try to really limit my oral directions as well. Part of conducting an investigation is defining it. I shouldn't take that opportunity away from the children.

There are times, however, when a teacher needs to direct the learning process: What assumptions are you making? Did you check your calculations? What did you do the last time you got stuck? What makes this question hard? Can you do anything about that? Prompting questions like these stimulate students to redirect their own thinking, unlike such judgmental comments as, You've made a mistake in your second step.

Questions are also useful for helping students clarify and extend their thinking: What do you mean by that? Can you think of another situation in which this would be true? Can you display your data in another way? Through such questions, the teacher can help students communicate their thoughts in a variety of forms and make connections between what is discovered and what is already known.

Finally, the teacher must model the same risk-taking and investigative thinking expected of students. Standards for performance must be high. Commitment to the task should be expected. At the same time, a variety of methods of communication must be respected in order to ensure that every student is able to demonstrate successfully what he or she knows.

CONCLUDING COMMENTS

The role of communication in learning and thinking mathematically cannot be overemphasized. Communication is central to students' formalization of mathematical concepts and strategies. Communication is essential to students' successful approach to, and solution of, mathematical explorations and investigations. Students must communicate with others to gain information; share thoughts and discoveries; brainstorm, evaluate, and sharpen ideas and plans; and convince others. For the teacher, communications are the vehicles for gaining insight into what students know and are able to do and for making important decisions about instruction and intervention. The explorations and investigations presented in this paper illustrate the richness of the educational experiences that can result from a variety of communications among students and between teachers and students. An understanding of the ways in which communication processes facilitate learning and thinking allows educators and curriculum developers to focus on the design of activities that foster and enhance these processes.

REFERENCES

Greenes, Carole, Linda Schulman, and Rika Spungin. "Stimulating Communication in Mathematics." *Arithmetic Teacher* 40 (October 1992): 78–82.

Greenes, Carole, George Immerzeel, Earl Ockenga, Linda Schulman, and Rika Spungin. *TOPS Problem Solving Card Deck* (CC, Grade 9). Palo Alto, Calif.: Dale Seymour Publications, 1980.

National Council of Teachers of Mathematics. *Curriculum and Evaluation Standards for School Mathematics.* Reston, Va.: National Council of Teachers of Mathematics, 1989.

———. *Professional Standards for Teaching Mathematics.* Reston, Va.: National Council of Teachers of Mathematics, 1991.

Secretary's Commission on Achieving Necessary Skills. *What Work Requires of Schools.* SCANS Report for America 2000. Washington, D.C.: U.S. Department of Labor, 1991.

21

Embedding Communication throughout the Curriculum

Harold L. Schoen

Diane L. Bean

Steven W. Ziebarth

THE main theme of this article is that students learn to communicate mathematically by being in an environment where such communication is a regular, natural, and valued occurrence. In such an environment, communication is prominent in instructional and assessment activities and the teacher gives regular formative feedback to students about their progress in communicating mathematical ideas. Students read, interpret, and conduct mathematical investigations in class. They frequently discuss, listen, and negotiate their mathematical ideas with other students individually, in small groups, and in a whole-class setting. Students write about mathematics and about their impressions and beliefs in group reports, personal journals, homework assignments, and assessment activities. They make oral reports in class that include communicating through graphs, words, equations, tables, and physical representations. Students are frequently involved in translating between the language of mathematics and that of technology. The context for all this communication is the curriculum itself. That is, the need and the opportunity for students to communicate with and about mathematics is embedded in the curriculum materials, the teaching strategies, the learning activities, and the assessment tasks.

In this paper, we present examples of such curriculum-embedded communication. The examples come from a high school mathematics curriculum now being developed by the Core-Plus Mathematics Project (CPMP). (See the endnote and Hirsch et al. [1995] for further information about this project.) The examples of students' work and teachers' reactions come from a yearlong pilot test of Course 1 in more than sixty ninth-grade classrooms. Several of these classes were a heterogeneous mix of the school's entire ninth grade, and nearly all classes contained students with a broader range of mathematical interests and past successes than is usually found in a single ninth-grade classroom.

COMMUNICATION THROUGHOUT CURRICULUM, INSTRUCTION, AND ASSESSMENT

The examples in this section show a variety of approaches for weaving communication into curriculum and assessment materials and some representative responses to the approaches from students and teachers. The examples are drawn from several of the content strands of the CPMP curriculum, which are algebra, data analysis, geometry, trigonometry, probability, and discrete mathematics.

Guidance for How to Communicate

The teacher-support materials offer a great deal of specific guidance for facilitating and assessing students' communication in cooperative groups and for managing and assessing students' products that require reading, writing, and oral communication skills. Guidance is also given directly in the student materials. This support is essential for students who are not used to communicating with and about mathematics. For example, rules for working in cooperative groups and descriptions of various roles that individual students play in these groups are stated and exemplified. In addition, students review, compare, and evaluate examples of written explanations of mathematical reasoning, as illustrated in figure 21.1.

Responses to parts (b), (c), and (d) will vary from group to group, but typical, appropriate answers would note that Paul's response was good because he described the ends of the distributions. Paul's response was incomplete because he did not describe the middle of the distribution or the gaps in it. Maria's response described the clustered nature of just a few values in the distribution, but she did not discuss its middle, gaps, or ends. Examples like this one provide models for students of what is expected when they are asked to write about their mathematical reasoning and of ways to judge different written explanations. Students coming from a traditional curriculum, we have learned, need this sort of guidance at first because they have rarely, if ever, been asked to write about their mathematical reasoning.

Communication in Group Investigations

Communication is a vital part of group investigations. Figure 21.2 is an example of a portion of a group investigation in which students are actively engaged in reasoning and communicating about simulation and probability. In the portion of the investigation given in figure 21.2, students work within groups and between groups as they design a simulation model to help them make sense of a real-life probabilistic situation.

(This example refers to a table giving the rankings
of fifteen popular music albums.)

When asked to describe the distribution of points for the top
fifteen albums, Paul and Maria first made a **stem-and-leaf
plot** like that shown at the right.

```
1 | 2 3 4 5 9
2 | 4 4 5 6
3 |
4 | 6
5 | 5
6 |
7 |
8 | 3
9 | 4
```

(a) In your group, discuss the organization of this stem-and-
leaf plot. Is this plot an accurate display of the data in the
chart? If not, explain how to correct it.

Paul's description: "Based on the stem-and-leaf plot, most
albums were ranked low. Two of the rat-
ings were high."

Maria's description: "You can see in the stem-and-leaf plot
that there were a whole bunch in the 10s and 20s. The
rest were spread out."

(b) What are the strengths and weaknesses of Paul's response?

(c) What are the strengths and weaknesses of Maria's response?

(d) As a group, write what you think would be a better response than either
Paul's or Maria's response.

Fig. 21.1. Students evaluate and revise written mathematical explanations

Almost any problem involving probability or an expected value can be solved
using simulation models. This final lesson of the unit provides three more such
problems to help you pull together the ideas you have developed and increase
your confidence in using simulation models.

1. About 10% of the adult population of the United States are African American.

   (a) Design a simulation model to determine the probability that a randomly
   selected jury of twelve people would have no African American members. Write out instructions for performing your simulation. Exchange instructions with another group.

   (b) Do the other group's instructions model the situation well? If necessary,
   modify the instructions and then conduct the simulation five times. Add
   your results to a copy of the frequency table below so that there is a
   total of 200 repetitions of the simulation.

Number of African Americans on the Jury	Frequency
0	56
1	73
2	45
3	16
4	4
5	1
Total	

   (c) Make a histogram of the results in your frequency table.

   (d) What is your estimate of the probability that a randomly selected jury of
   twelve people would have no African American members? Justify your
   answer.

Fig. 21.2. Communication embedded in a group investigation on simulation and probability

Communication in Checkpoints

At the end of each small-group investigation and at the end of each unit is a feature called a "checkpoint." A checkpoint is a small-group activity followed by a whole-class discussion in which representatives of each group communicate what transpired in their group by responding to the synthesizing questions in the checkpoint. The checkpoint in figure 21.3 comes at the end of a unit on network optimization. The main topics of the unit (minimal spanning trees, the traveling-salesperson problem, and shortest paths) can be inferred from the questions in the checkpoint.

√ Checkpoint

(a) The title of the first lesson of this unit is "Finding the Best Networks." "Finding the best" is an important theme throughout the entire unit. Describe three examples from the unit for which you "found the best." In each example tell which graph model you used to solve the problem.

(b) Describe one similarity between minimal spanning trees and shortest paths. Describe one difference.

(c) Describe one similarity between minimal spanning-tree problems and the traveling-salesperson problem. Describe one difference.

(d) Describe one similarity between the traveling-salesperson problem and the shortest-path problems. Describe one difference.

**Be prepared to share your group's examples and
descriptions with the class.**

Fig. 21.3. Communication embedded in a checkpoint in the network optimization unit

In general, a checkpoint generates a great deal of communication, both within students' groups and with the entire class. The whole-class discussion focuses on the commonalties and differences in the groups' methods and results and ordinarily culminates in a summary of what was learned in the preceding investigation or unit. Students learn a great deal about mathematics and how to communicate it from hearing and discussing the groups' reports. The teacher learns how groups are working together and what they understand and can communicate about the mathematics. The teacher can use this information to adjust the group activities as needed, so a checkpoint also serves as a curriculum-embedded assessment.

Communication in Written Reflections

Many students keep journals in which they write regularly concerning their personal impressions, learning difficulties, interests, and so forth. Such written reflection is further encouraged by reflecting tasks at the end of each lesson. The purpose of a reflecting task is usually to prompt a written journal response as part of a homework assignment.

The reflecting task in figure 21.4 appears near the end of a unit on exponential models. In this unit, students explore many real-world situations that involve exponential growth and decay. This reflecting exercise prompts students to think and write about their beliefs concerning the relevance and importance of the situations they have just investigated.

> Which example of an exponential decay pattern—the Sierpinski carpets, the metabolizing of drugs in the body, the bouncing of a golf ball, or the decay of radioactive chemicals—seems to you to be the most interesting or important example of exponential decay? Why?

Fig. 21.4. A reflecting task concerning examples of exponential decay

Communication in Quizzes and Examinations

Assessments are aligned with the curriculum and the instructional approach by design, so it is often difficult to distinguish an assessment from an instructional activity. In fact, group investigations and checkpoint discussions serve simultaneously as instruction and assessment. Sometimes, though, teachers want to give students—individually or in small groups—fairly well defined tasks with the express purpose of assessing their levels of understanding at a given time. Assessment tasks of different lengths that are administered differently are included for this purpose. For individual in-class assessments, twenty-minute quizzes for each lesson and fifty-minute end-of-unit assessments are provided. Questions to take home overnight are also furnished at the end of each unit; these are usually appropriate for individuals or pairs of students. Finally, one-to-two-week extended projects that apply the main ideas of the unit holistically can be completed by pairs or small groups of students. In-class and take-home cumulative assessments are also provided for each set of several logically connected units and for the entire course.

All assessments, even the quizzes, require students to construct their responses, show and justify their work, and explain their reasoning. The in-class assessment task in figure 21.5 is from a cumulative assessment that follows three units in which graphing relations and functions play an important role. It requires students to show their proficiency in two important aspects of mathematical communication: (1) translating between a verbal description of a familiar real-world situation and a graph of the relationship of two important variables in that situation and (2) writing an explanation of the reasoning that led to the graph. Responses by students to this task furnish a clear window into the variety of ways in which they think about the connection between these two representations. Three examples of fairly representative responses are given in figure 21.6.

A little girl is swinging on a playground swing. She is being pushed higher and higher by her brother.

(a) Draw a graph of the height of the little girl above the ground as a function of time. You will not be able to give exact heights or times, but your graph should be approximately the right shape and in the right position.

(b) Explain how and why the little girl's height above the ground changes over time in the way your graph suggests.

Fig. 21.5. An assessment task that requires constructing a graph of a real-world function

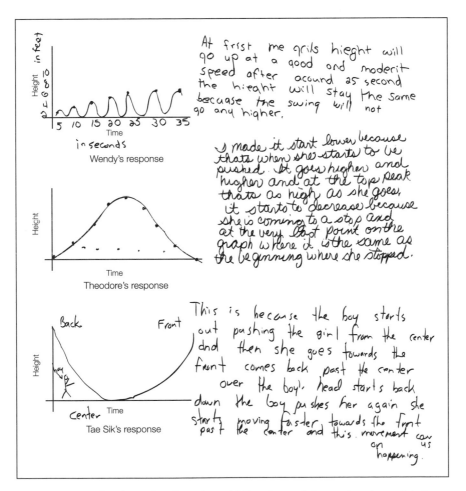

At frist me grils hieght will go up at a good and moderit speed after acound 25 second the hieght will stay the same becuase the swing will not go any higher.

Wendy's response

I made it start lower because thats when she starts to be pushed. It goes higher and higher and at the top peak thats as high as she goes, it starts to decrease because she is coming to a stop and at the very last point on the graph where it is the same as the beginning where she stopped.

Theodore's response

This is because the boy starts out pushing the girl from the center and then she goes towards the front comes back past the center over the boy's head starts back down the boy pushes her again she starts moving faster towards the front past the center and this movement can on us happening.

Tae Sik's response

Fig. 21.6. Students' graphs of a real-world function and the explanations of their reasoning

Wendy's graph is more or less what we had in mind when we wrote the task. It captures the up-and-down motion of the swing and the increase in the peaks over time. Her explanation focuses on the increasing height, and she even recognized and articulated that the height could not increase indefinitely. She does not explain why she put numbers on the axes or why those particular numbers, even though we had indirectly discouraged them by saying, "You will not be able to give exact heights or times."

At first glance, it may appear that Theodore has drawn one period of Wendy's graph and that their thinking might be similar. A closer look reveals that Theodore marked maximum points and minimum points of each arc of the swing with dots but then joined the consecutive maximum points only with a smooth curve. He also decided, correctly, that the swing would eventually have to stop and, incorrectly, that the time-and-height graph of the entire swinging episode would have a symmetric, parabolic shape.

Like Wendy and Theodore, most students constructed graphs that showed that they were in some way relating time and height. A few, like Tae Sik, confused the graph with the real-world situation and actually sketched a boy pushing a little girl in a swing where the curve is the path of the little girl relative to the ground (the horizontal axis). Tae Sik, in his explanation, clearly articulated what he was doing. In fact, most students explained their thinking on this task quite clearly, a striking improvement over their muddled explanations (and often nonexplanations) on an open-ended, preassessment task administered about three months before this assessment was completed.

This task and the accompanying student responses also illustrate the richness of the information that can be gained from assessments that require students to construct and explain their responses. In all three examples, the student's graph gave some information about his or her thinking, and the explanation added a great deal more. In addition, the task gives students the message that reasoning and explaining are an important part of mathematics and, further, that their own mathematical reasoning is important enough for their teacher to assess it.

Communication in Extended Projects

Extended projects afford the opportunity for pairs or small groups of students to work together to organize the important concepts from the recently completed unit or units and to extend these ideas into other areas of interest to them. This kind of assessment also allows students whose learning styles and dispositions put them at a disadvantage on in-class examinations to demonstrate their holistic grasp of some of the major ideas in the units in a culminating activity. These projects typically require a week or two to complete, and the teacher provides periodic feedback as the students develop the end product. Figure 21.7 shows an extended project that follows a unit on geometric form and its function.

Unit 6 Project Title: **Linkages, Gears, and Motion**

Purpose: This unit described various linkages and the geometric properties that allow them to function. It also introduced some of the interesting and useful properties of circular motion, including its role in pulley and gear systems. In this project, you will examine in detail another example of your choice and describe the geometric properties that it uses.

Directions

1. Choose a linkage, gear system, carnival ride, or other mechanism that is of interest to you; study it and describe the geometric properties that make it work or the nature of its motion. Examples include, but are not limited to, the following:

   - A linkage that is different from any in the unit for a folding or reclining chair, a jack, a door system, a dump truck, and so forth
   - A multigear bicycle or other vehicle different from those considered in the unit
   - A carnival ride in which riders go in circles while the floor fluctuates up and down or a double Ferris wheel that has a rotating Ferris wheel mounted on either end of a giant bar that also goes around.
   - The path of a point on a wheel of an off-center axle, for example, a clown's bicycle.

2. If possible, take measurements and gather data from the actual mechanism that you choose in part 1. For example, measure the important parts of a linkage you choose, sketch the linkage, and give all measurements indicating which measurements vary, what geometric shape is used, and so forth. For a bicycle or other gear system, measure or count the teeth in each pulley or sprocket. Actually test the properties of the gear system by experimenting and measuring appropriate distances or speeds. If the mechanism is not available (such as a carnival ride), make a miniature model of it and gather data about the nature of its motion using the model.

3. Write a report in which you sketch—in several positions if appropriate—and describe the mechanism that you have studied. Describe exactly what you did to study the properties and why. Include the data that you gathered, and model and explain the mathematical properties or relations that account for the data. When you are not sure of a mathematical model, at least present your data with tables and graphs and do your best to explain it verbally.

Fig. 21.7. An extended project emphasizing written communication

It is important that students be given an adequate amount of time and regular, intermediate encouragement and feedback. We suggest that students should have several days to a week to decide on the focus of their project, another few days to study it or develop a model, several more days to submit the first draft of their report, and at least a week after that to prepare the final report. In between, allow some class time for students to ask questions and to work on the reports. If time permits, an oral report to the rest of the class may be worthwhile, too, since different groups will be studying different linkages, gear systems, and so forth.

If the teacher decides to assign a grade to the report, the following guidelines, which emphasize communication skills as well as mathematical accuracy, are suggested:

1. Adherence to directions for format   20%
2. Choice of interesting mechanism and appropriate methods   20%
3. Quality of sketches, measurements, or tests to illustrate
   geometric properties   30%
4. Quality of explanation of why and how the mechanism works   30%

Other extended projects involve students in such activities as taking photographs and making posters that illustrate certain geometric shapes and symmetries in human and natural design; writing mathematically supported arguments to defend a position on an issue of current and perhaps local interest, such as who is the better of two athletes or which television network has the best lineup of shows; doing library research projects involving gathering data from reference books about such important issues as population growth; conducting and reporting the results of surveys of public opinion; designing a plan for a school or community project, such as paper recycling or a school dance, that requires gathering data and analyzing them with statistical, algebraic, geometric, or discrete mathematics methods; and designing and conducting a simulation model of a situation of interest to students. Communication is an important part of every extended project, since each requires a written—and sometimes oral or pictorial—student product that communicates the way in which mathematics was applied, why it was applied in that way, and what the results were.

CONCLUDING COMMENTS

The main theme of this article is that students learn to communicate mathematically by being in an environment where such communication is a regular, natural, and valued occurrence. The foregoing examples are meant to clarify how this theme translates into practice in one curriculum.

Unfortunately, most students' experiences have led them to believe that the valued and assessed outcomes in mathematics are mainly proficiency in, and routine application of, symbol manipulation. If you begin to embed communication in your mathematics curriculum in some of the ways suggested in this paper, be prepared for an initial period of resistance from your students. Pilot teachers have uniformly met with resistance at first. Students were not accustomed to reading, interpreting, and investigating mathematical material, discussing mathematics in groups, and writing about their work and their reactions to it. Many students felt at first that this was not mathematics, that is, procedures for manipulating symbols.

However, students using this curriculum soon begin to realize that communication, teamwork, real-world contexts, multiple representations, investigation, and thinking are valuable parts of mathematics, learning, and assessment. In fact, as students have commented, "working in groups has helped me ... because I can understand things better when people my age try to explain things because they don't use words I don't know" and "I can explain how I came up with my answer and I understand and I can help someone else to understand."

The Core-Plus Mathematics Project (CPMP), funded by a five-year grant from the National Science Foundation, has been developing and field-testing an innovative, three-year, high school mathematical-sciences core curriculum for all students, plus a fourth-year course that continues the preparation of students for college mathematics. The four-year program emphasizes mathematical modeling and student investigation. Interwoven strands of algebra, geometry, trigonometry, probability, statistics, and discrete mathematics are developed each year through realistic applications. Units are connected by common themes of data, shape, change, and representation; by such common topics as matrices, symmetry, curve-fitting, and recursion; and by mathematical habits of mind. The investigation of real-life situations leads to the invention and reinvention of important mathematics that makes sense to students and in turn enables them to make sense of new situations and problems.

Codirectors of the Core-Plus Mathematics Project are Christian Hirsch (Western Michigan University), Arthur Coxford (University of Michigan), James Fey (University of Maryland), and Harold Schoen (University of Iowa). Other principal curriculum developers are Gail Burrill (University of Wisconsin), Eric Hart (Western Michigan University), and Ann Watkins (California State University—Northridge).

This project is supported, in part, by National Science Foundation Grant No. MDR-9255257. The opinions expressed are those of the authors and not necessarily those of the foundation.

Reference

Hirsch, Christian R., Arthur F. Coxford, James T. Fey, and Harold L. Schoen. "Teaching Sensible Mathematics in Sense-Making Ways with the CPMP." *Mathematics Teacher* 88 (November 1995): 694–700.

22

Children, Teach Your Parents Well:
Communication in Mathematics between Home and School

Amy Hart

Mitzi Smyth

Kate Vetter

Eric Hart

Every mathematics classroom is part of a larger community—the school-home community. It is essential to have ongoing communication of and about mathematics within this community. In fact, a recent analysis of successful Japanese mathematics education programs (Reys and Reys 1995) points out that an important factor contributing to students' success in mathematics is the cooperative effort between parents and school to help the child. This article presents one way to foster communication between the home and the classroom: have children teach mathematics to their parents or guardians.

A VARIETY OF APPROACHES

There are many ways to get children involved in teaching mathematics to their parents—ranging from quick puzzles to brief explanations to an actual lesson. Children can have fun challenging their parents with puzzles they have already solved in mathematics class. Parents of fourth graders in our area have spent many spare moments trying to figure out puzzles like this: Make the equation $3+3+3=346$ true by drawing one straight line, horizontally, vertically, or diagonally (Cook 1993).

A more involved child-parent interaction might consist of having children briefly explain to a parent something they have recently learned in mathematics class. This interaction can be formalized by having the parent complete a simple form like the one in figure 22.1.

Homework Assignment Form

To be signed by parent and returned by student on Monday morning

My child_____ taught me

_____.

The lesson was well taught and well learned.

_____

Parent's Signature

Fig 22.1

Because of everyone's busy schedule, it is often possible to have only brief child-parent mathematics interactions, like the puzzle or homework-form examples. But now and then it is well worth the time for parents and children to have a more in-depth interaction, one where the child actually teaches a short mathematics lesson.

CHILDREN TEACHING PARENTS

It takes some time and preparation for children to productively teach a short lesson to their parents. We present here one detailed example that shows how the process has worked in our area. The children learn a lesson, then they learn and practice how to teach it, and finally they take the materials home and teach their parents.

The Lesson: Areas of Parallelograms

The general concept in this lesson is area. In particular, fourth graders investigate the relationship between the area of a rectangle and the area of a nonrectangular parallelogram. The children spend one class period doing the following activity.

They begin by cutting out a rectangle and a nonrectangular parallelogram that have the same length and height from two-centimeter graph paper (although the children are not alerted that the lengths and heights are equal). They are then asked to "play around" for a while and see what they can find out about the areas of the two figures. After a brief time, the teacher instructs them to cut off one corner of the parallelogram to see if this helps them figure out how the two areas are related. With appropriate guidance, the children eventually see that the cut-off corner can be repositioned on the other side of the parallelogram. They thus find that the two figures have equal areas.

When they begin to think about formulas for determining area, they see that the rectangle-area formula, length × width, does not quite work, since the "width" of the parallelogram is slanted. They continue working with their cutouts and ultimately conclude that the area of the parallelogram is length × *height*. Figure 22.2 shows children at work on this activity.

Fig. 22.2

Preparing to Teach the Lesson

Teachers should be forewarned that it can take up to two class sessions to get the children prepared to teach the lesson to their parents. But the benefits in the students' learning and in school-home communication make the time a good investment, if such an activity is done once in a while with important and appropriate topics.

First the children teach each other in pairs, not just explaining with existing cutout figures but teaching the lesson from scratch. They find out that teaching is not as easy as it looks! After this experience, the teacher and students reflect on the teaching and discuss some ideas for making it easier and more effective. Led by the teacher, they work together to get an initial lesson plan. Then the children practice the lesson plan by teaching each other in pairs once again. Finally, the class creates a three-part "Parent's Homework" assignment and a simple take-home lesson plan, shown in figure 22.3.

How Did the Parents Do?

The children went home armed with a letter from the teacher explaining the child-parent teaching idea, the Parent's Homework activity sheet,

Parent's Homework	Lesson Plan (directions and hints)
1. What does this tell you about the area of the rectangle and the parallelogram?	Draw, cut, play around, compare areas.
2. What does this tell you about the rule for finding the area of a parallelogram?	Compare widths, $l \times w$ works?, find rule.
3. Find examples of parallelograms outside or in your house or apartment.	Make list.

Fig. 22.3

the lesson plan, and graph paper. Figure 22.4 shows one parent hard at work under the guidance of a wise teacher. Almost all the parents completed the activity and turned their homework in on time. Here are some of their responses.

Task 1—What does this tell you about the area of the rectangle and the parallelogram?

- "The areas of both are the same even though the figures appear to be different sizes."
- "They have the same area."
- "Both the rectangle and the parallelogram have the same area— 15 squares or 30 centimeters. I multiplied length times width

Fig. 22.4

on the rectangle and superimposed the parallelogram over the rectangle to visually ascertain their equal area."

Task 2—What does this tell you about the rule for finding the area of a parallelogram?

- "The area of a parallelogram is not calculated by multiplying the width of one side by the length of the other."
- "The area is the total number of squares across (length) multiplied by the total number of squares from top to bottom (height)."
- "You multiply the length of the base times the height of the middle section."
- "You find the area of a rectangle and then find the area of a triangle and then add them."
- "Finding the area of a parallelogram is different from finding the area of rectangles. You multiply length by height instead of length by width."

Task 3—Find examples of parallelograms outside or in your house or apartment.

- "There are many examples of rectangles and squares, which are parallelograms, but only a few of the general shape."
- "Owen's blocks"
- "Carved wood on stairway"
- "Shelf behind the stove, bottom of shelf of wicker table, stained-glass window design, pieces of quilt, pieces of tile flooring, lattice work on fence"

THE AFTERMATH

The children enjoyed this activity very much. The following day they were all clamoring to report how their mom or dad learned the lesson. The teacher asked them to reflect briefly on the activity by writing the answers to these three questions in their journals: What was the easiest part of the teaching activity? What was the hardest part? What did you learn? Some responses are shown in figure 22.5.

One Child's Viewpoint

One child wrote the following assessment of the activity:

This was really fun and my dad was a good student. We worked for about half an hour. He didn't get how to draw the parallelogram at first. I explained again and then he got it. He learned all the other stuff really well.

Sometimes in math lab when I don't get what's happening I just say, oh well, but when you have to teach you have to really get it! When I was first

"What did you learn?"

I learned how to teach it
in school when Miss Vetter
explained it and when I
taught it to my friends

My Dad was teaching me at home
you can do L x W on a P-gram

I learnd I don't wont to be a
techer

I Learned I can teach to!

"What was the easiest part?"

telling your parent how to draw
the Rectangle

to cut out the shapes

everything was easy
because she new it from
her own experieness

"What was the hardest part?"

Leaning how to teach
the math thing.

the hardest part about teaching
is teaching your parent about
the length x hight part.

Fig. 22.5

learning I didn't understand the formula part at all. It was the second time that I taught the lesson to my friends in class that I finally got it. I would like to teach my dad again but I hope I understand the math before the last minute.

It was a lot of work to learn the math and learn to teach but we had fun and I felt good to be able to teach it to my dad.

CONCLUDING COMMENTS

Children, parents, and teachers thought this was a worthwhile activity. However, it does take time, so you need to be sure that the mathematical concepts involved are important. In this activity, children learned about area and how to generate and make sense of symbolic formulas. Also, the context should be concrete, allowing the children and parents to do and show, not just explain. To make it more interesting and fun, you should try to find topics that the parents might not be so familiar with; yet at the same time the topics should be nonthreatening and they should send home the right message about what you think is important in mathematics and how children learn it.

By teaching their friends and parents, the children internalized, reflected, and really learned. But in addition to learning mathematics, there were several other positive outcomes. Thinking about how to teach gave the children an opportunity to develop important metacognitive skills, as well as organizational and communication skills. Strengthening the school-home link proved to be one of the most positive outcomes. Mathematics is an area where many parents feel isolated from their children. Activities like these are needed to foster stronger communication between home and school, so that learning can be enhanced. As both research and experience clearly show, parental involvement in education leads to improved learning and teaching. Finally, we found, beyond doubt, that children *can* "teach their parents well"!

REFERENCES

Cook, Marcy. *Math Starters and Stumpers*. Balboa Island, Calif.: Marcy Cook, 1993.

Reys, Barbara J., and Robert E. Reys. "Japanese Mathematics Education: What Makes It Work?" *Teaching Children Mathematics* 8 (April 1995): 474–75.

23

Communicating about Alternative Assessment beyond the Mathematics Classroom

Patricia Ann Kenney

Cathy G. Schloemer

Ralph W. Cain

Mathematics teachers have the responsibility of communicating about alternative assessment methods to a variety of interested parties including students, parents, guardians, and school administrators. At first, communicating effectively about alternative assessment may prove to be a different, and somewhat difficult, undertaking for most teachers of mathematics. Also, given the grading practices that currently exist in most schools, teachers must grapple not only with how to transform fairly and accurately the results from alternative assessment into a letter grade, a numerical grade, or a written report of student progress (e.g., checklists, student profiles) but also with how to explain—and sometimes to justify—their grading practices to others.

It is important that teachers develop effective ways to communicate information about alternative assessment practices not only within the mathematics classroom, as they share their expectations for instruction and assessment with students, but also beyond the mathematics classroom. In fact, providing information about the mathematics assessment process is embodied in the Openness Standard—"Assessment should be an open process"—of the *Assessment Standards for School Mathematics* (NCTM 1995), the document that presents the assessment vision of the National Council of Teachers of Mathematics in its efforts to guide reform in the teaching and learning of mathematics. The purposes of this article are (1) to investigate ways in which teachers can showcase the benefits of using alternative assessment methods to parents, guardians, school administrators, and other stakeholders and (2) to suggest how teachers can respond to the questions or challenges involved in using results from such assessments as the basis for making decisions about further learning and in the summative evaluation of students' achievement in mathematics.

COMMUNICATING ABOUT THE BENEFITS OF USING ALTERNATIVE ASSESSMENT METHODS

Convinced that using alternative assessment methods in their classrooms results in better teaching and learning for students, teachers often find themselves in the position of having to explain to others *why* these methods are better than other, more familiar kinds of assessment. For most teachers, especially those considering a radical change from what they have done before or what is customary in their school, talking with the principal or school administrator about changes in assessment practice is a necessary first step. Sharing a concrete plan that includes the kinds of assessment methods to be used (e.g., observations, journal writing, portfolios), the benefits that come with using these methods (e.g., information about the process as well as the product, encouragement of reflective thinking through explanation), and the ways in which students' work will be evaluated (e.g., checklists, scoring guides) will likely result in gaining the administrator's support. And administrator support often paves the way for a collaboration in which teachers share assessment methods with one another. In addition to communicating to school administrators, mathematics teachers also need to establish and foster a classroom environment in which the students enter the assessment process well informed about what they need to know, how they will be expected to demonstrate it, and what the consequences of the assessment will be.

Certainly, parents or guardians need to have information about the ways in which their children's performance in mathematics will be evaluated during the course of the school year. One way to bring these people into the information loop is to share, in writing, essential parts of the classroom assessment plan developed for the school administrator. Asking parents to sign and return a portion of the page and including a telephone number at which the teacher can be reached further encourage communication between teachers and parents about assessment issues.

So far, the suggestions for facilitating communication about the benefits of alternative assessment have been quite general. However, some examples of how to communicate about specific assessment techniques can be helpful. Although there are many forms of classroom assessment techniques and many interested parties with whom to communicate about these assessments, we present only a few representative examples here. For some examples we also include a brief summary of the assessment method, but the main focus will be on ways to communicate about the benefits afforded by a particular alternative assessment method. More information about assessment in mathematics and related issues is available in publications (e.g., Charles, Lester, and O'Daffer 1987; Stenmark 1991; Webb 1993), in reports of state assessment programs (e.g., Mumme 1990; Pandey 1991; Pennsylvania Department of Education 1993; Vermont Department of Education 1991), and in the *Assessment Standards for School Mathematics* (NCTM 1995).

Example: Communicating about the Benefits of Using Open-Ended Tasks

In contemporary discussions about assessment reform, open-ended mathematics tasks in which students must show their work and explain their reasoning have been showcased as an effective way to assess students' mathematical power. Multiple-choice questions, conversely, have been criticized as being too limited to assess many valued aspects of mathematical performance. Yet it is likely that the traditional multiple-choice format is very familiar to most parents or guardians. Capitalizing on that familiarity can provide teachers with a unique opportunity to communicate about the benefits of using alternative assessments through a comparison of students' performance on a multiple-choice question and an open-ended version of that same question. In the example that follows, we suggest a way to point out the benefits of a particular open-ended task by comparing it to its multiple-choice counterpart. Although we present only one example, there are many multiple-choice questions that can be transformed in this way and then used in the classroom and as a part of a Family Night presentation.

The multiple-choice question in figure 23.1 is from the 1986 National Assessment of Educational Progress (NAEP) in mathematics. It assesses students' understanding of least common multiple—an important concept in number theory—but its multiple-choice format places the focus on obtaining the single, correct answer. Besides learning that some students were able to select the correct answer through the use of some unknown process (or by guessing) and that other students selected the incorrect answer through the use of a flawed process, teachers gain little direct information about their students' mathematical knowledge from this question in its present format.

But what if that multiple-choice question were transformed into an open-ended task? A modified version of that question (fig. 23.2) can be

---

When the items in a box are put in groups of 3 or 5 or 6, there is always 1 item left over. How many items are in the box if there are fewer than 50?

  A) 16   B) 29   C) 30   D) 31

Fig. 23.1. Multiple-choice question

When the items in a box are put in groups of 3 or 5 or 6, there is always 1 item left over.

How many items could be on the box?

Show your work.

Sample A

Sample B

$$3, 6, 9, 12, 15, 18, 21, 24, 27, \boxed{30}$$

$$5, 10, 15, 20, 25, \boxed{30}$$

$$6, 12, 18, 24, \boxed{30}$$

30 is common   add 1

$$\underline{\underline{31}}$$

Sample C

$$3\overline{)31} \quad \begin{array}{r} 10 \\ \hline 31 \\ 30 \\ \hline 1 \text{ left} \end{array}$$

$$5\overline{)31} \quad \begin{array}{r} 6 \\ \hline 31 \\ 30 \\ \hline 1 \text{ left} \end{array}$$

$$6\overline{)31} \quad \begin{array}{r} 5 \\ \hline 31 \\ 30 \\ \hline 1 \text{ left} \end{array} \qquad 31$$

$$3\overline{)61} \quad \begin{array}{r} 20 \\ \hline 61 \\ 60 \\ \hline 1 \text{ left} \end{array}$$

$$5\overline{)61} \quad \begin{array}{r} 12 \\ \hline 61 \\ 5 \\ \hline 1\ 6 \\ \hline 1 \text{ left} \end{array}$$

$$6\overline{)61} \quad \begin{array}{r} 10 \\ \hline 61 \\ 60 \\ \hline 1 \text{ left} \end{array} \qquad 61 \text{ works, too.}$$

Fig. 23.2. Open-ended question and sample responses

formed by deleting the choices and changing the wording accordingly. The open-ended version still assesses the same important mathematical concept as the multiple-choice question, but it does so in a way that encourages students to write down the process they used and allows for multiple correct answers (see sample responses in fig. 23.2). In addition to finding out how some students arrived at one or several of the correct answers, teachers using the open-ended form of this task also obtain important information about why other students failed to obtain a correct response and about which misconceptions were most common. On the basis of students' responses to open-ended tasks, teachers can gain important insights into students' thinking and reasoning that can be used in making instructional decisions. Also, by promoting a variety of answers and strategies, open-ended tasks encourage the adoption of more holistic scoring schemes based on partial credit. There are many sources of scoring guides for these kinds of tasks, and teachers can adapt a scoring guide developed for a state- or national-level assessment and then use it as the basis for their own evaluation. For example, the four-level scoring scheme (minimal, partial, satisfactory, extended) used to evaluate open-ended questions from the 1992 NAEP mathematics assessment serve such a role. (A complete version of the scoring guide can be found in Dossey, Mullis, and Jones [1993, p. 89].)

Once teachers have samples of their own students' work on open-ended mathematics tasks, they can use these samples to communicate with parents, guardians, and others about the benefits of using such tasks to assess students' understandings and misunderstandings. For example, during a school open house a teacher presents the multiple-choice question and the open-ended task with its scoring guide to those in attendance, and then shows examples of students' work. Parents and guardians can then see first hand the kinds of information afforded by the open format that are just not available from the closed format. In the case of the open-ended task used here as an example (see fig. 23.2), parents and guardians can see the kinds of strategies that students employed to solve the problem and the kinds of errors they made, such as those involving minor computational mistakes or adding the numbers mentioned in the problem and getting an answer of 17. Sharing the scoring guide for the open-ended task facilitates communication with others about the criteria used to evaluate students' performance on open-ended mathematics tasks.

Example: Communicating about the Benefits of Using Portfolios

Open-ended tasks and other methods of alternative assessment individually offer a snapshot of students' achievement in mathematics. Portfolios are an important way to integrate results from a variety of alternative assessment methods to demonstrate tangibly and over time what students are learning. As documented in other publications (e.g., Mumme 1990; Stenmark 1991), the contents of a portfolio can include a variety of formats such as teacher-completed checklists, notes from an interview, papers showing a student's corrections of errors or misconceptions, reports of group projects, and entries from a student's journal.

Teachers who incorporate portfolios into their classroom assessment schemes have a natural way to communicate to others the benefits of this method by using the activities included in a student's mathematics portfolio. The following example is from a first grader's portfolio. (The example is from a portfolio assessment activity used by Sue Coburn, a first-grade teacher at Kerr Elementary School, Pittsburgh, Pa.) For this activity, each student used a geoboard and rubber bands to make a pattern and was instructed to copy the pattern onto dot paper. This same activity was done a number of times over a two-month period, and the teacher used a rubber stamp to ink the date on the front of each paper. The drawings shown here (fig. 23.3) were done by the same student: one drawing on 6 October and the other on 10 December. Comparing the two drawings clearly shows that the student became much more elaborate in creating patterns over time and also became much more perceptive of kinds of number patterns (e.g., the numerical pattern [1, 2, 3, 4, 5, 4, 3, 2, 1] in the 10 December drawing).

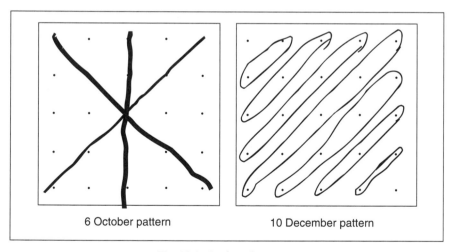

6 October pattern   10 December pattern

Fig. 23.3. Geoboard patterns

Through the use of tasks like the one described above and other tasks appropriate for the particular grade level and curriculum, a portfolio containing samples of a student's work shows in a dramatic way how that student's level of mathematical power changes over time. Portfolios are a tangible way to communicate with others about what students are learning about mathematics. A parent-teacher conference is one setting that is particularly appropriate to showcase the benefits of using portfolios as an assessment method. To encourage parents or guardians to attend a conference to discuss the portfolios, the teacher can send a note home a month or so before the conference containing a message such as, "I look forward to sharing your child's accomplishments with you at our May conference." Later, when the teacher sends home a note to set up a specific appointment with the parents, the teacher can emphasize that because they will be *viewing* their child's materials, a telephone conference is not as productive an alternative. At conference time, the portfolio contains evidence that facilitates communication between teachers and parents or guardians about students' progress that is probably not readily available from report card grades or other forms of summative reporting and that may not deal exclusively with mathematics. For example, through the two geoboard drawings (see fig. 23.3) and other available information, the teacher can point out the child's progress in hand-eye coordination (e.g., the accuracy in the 10 December drawing as opposed to the missed dot in the lower right-hand corner of the 6 October drawing).

COMMUNICATING ABOUT THE DIFFERENCES BETWEEN ASSESSMENT AND GRADING

We open this section with two anecdotes. The first anecdote involves a group of middle school teachers attending a workshop on evaluating responses to open-ended mathematics tasks. The teachers who participated in the workshop evaluated a set of student responses using a focused, holistic scoring scheme with five score levels (level 0 to level 4), and they also looked at responses for strategies, error patterns, and modes of representation. Despite the workshop's focus on alternative assessment and on the benefits from using open-ended tasks, the very first comment during the question-and-answer period was, "This is all well and good, but in my school I *have* to give letter grades every six weeks. Can I assume that a '4' is a 90, a '3' is an 80, and so on, with a '0' as a failing grade?"

The second anecdote involves a high school mathematics teacher who throughout the semester had tried to establish an open assessment process in her classroom in which students would know what they had to learn and how they would be expected to demonstrate that learning. Yet, one student was puzzled when she encountered this question on a trigonometry test: "Explain the following statement and illustrate with several examples: 'As x increases from 0 to 1, $1/x$ decreases from an undefined value to 1.'" Despite the fact that this issue had been discussed at length in class, the student said, "You've never asked us to do this in class. Why are you doing it now on the test?"

Both of these anecdotes are examples of how miscommunications can occur—even when teachers make a conscious effort to communicate their expectations to students and even when teachers themselves are encouraged to examine students' responses beyond the correct answer. And if such misunderstandings exist among those within the mathematics classroom, it is certainly likely that miscommunications will also occur between teachers and other stakeholders who do not come in daily contact with the mathematics classroom. In particular, as the anecdotes presented above suggest, the tension between alternative assessments and traditional grading practices is one area for potential miscommunication. Although open-ended tasks, portfolios, and other alternative assessment practices can furnish teachers with a wealth of useful information, for many students and their parents or guardians the most important outcome of any classroom assessment procedure is the contribution it makes to the summative evaluation of achievement; that is, *grades.* Grading looms large in the life of classroom teachers, especially at the middle school and secondary levels, yet teachers have little—if any!—say in the establishment of grading practices. This decision is most often made at the district level and has traditionally involved reporting student achievement in mathematics as an accumulated average score (e.g., 72) or by using letter grades to represent ranges of scores (e.g., 80 to 89 = B). Even when the summative evaluation method is based on a checklist format including topics like addition facts, counting, and geometry, teachers still must aggregate results from alternative assessments and translate them into marks in categories like "satisfactory," "needs improvement," or some other indicator of a student's level of achievement.

As a consequence, mathematics teachers find themselves struggling with the tension between implementing alternative assessment methods—the purpose of which is primarily formative and intended to establish a communication link between students and teachers—and fulfilling district grading requirements based on specified, rigid ways to report student achievement (Seeley 1994). In her analysis of the *Curriculum and Evaluation Standards for School Mathematics* (NCTM 1989), Lambdin (1993, pp. 12–13) speaks to this tension:

> [T]he most formidable impediment to innovative assessment techniques may be tradition. Educational assessment procedures that have been in place for decades are difficult to change. Tests and letter grades are well established as methods for evaluating and reporting students' achievements in mathematics.... Even if teachers are convinced of the benefits of using more innovative methods to evaluate their students, they are unlikely to succeed unless their supervisors, students, parents—and even their fellow teachers—understand and support their break with tradition.

Although the reform movement in education is currently questioning the validity of current grading practices in all school subjects, implementing changes will take time. In the interim, to ensure the kind of "success" that Lambdin describes, mathematics teachers need to communicate clearly about their grading practices and, in particular, about how grades are derived from alternative assessment results.

Teachers, who share with parents, guardians, and school administrators their partial-credit scoring guides, checklists for interviews, plans for portfolio evaluations, and other schemes for evaluating the results from alternative assessments have already laid the groundwork necessary for communicating about their grading practices. One experienced high school mathematics teacher advocates what she calls the "first strike" method: that is, communicating with parents, guardians, school administrators, and the students about alternative assessment practices and evaluation techniques well before the first report cards are sent home. For example, attaching the scoring guide used with a particular task to the response supplies an informal report to the student and the parents or guardians. The scoring guide also demonstrates that the teacher's evaluation was essentially objective, or "scorer free"—that is, a score that is a function of the response and not a subjective evaluation. In the example of long-term assessment methods such as portfolios, it is important for teachers to decide in advance whether to use portfolio assessment as a way to substitute or displace some dependence on grades (e.g., sharing the contents with parents during a conference as a way to communicate what their child is learning) or as a summative body of information that can become part of the grading scheme (e.g., assigning the portfolio a percentage [20%; 40%] of the grade) (Mumme 1990).

CONCLUDING COMMENTS

Communicating effectively about the role of alternative assessment practice in the mathematics classroom will become an increasingly important role for teachers, especially as the vision of the NCTM's *Curriculum and Evaluation Standards for School Mathematics, Professional Standards for Teaching Mathematics* (1991), and *Assessment Standards for School Mathematics* permeates the teaching and learning of mathematics. This article has suggested some ways for teachers to promote the idea that alternative assessment methods offer an expanded view of students' mathematical power and that results from alternative assessments can be effectively incorporated into traditional grading practices as well as into the assessment practices of the future. Effective communication about alternative assessment and its role in describing students' mathematical power can make parents, guardians, school administrators, and other stakeholders more aware of the benefits of these methods and further the movement toward evaluation models that retain the breadth and depth of evidence gathered from multiple sources.

REFERENCES

Charles, Randall, Frank Lester, and Phares O'Daffer. *How to Evaluate Progress in Problem Solving.* Reston, Va.: National Council of Teachers of Mathematics, 1987.

Dossey, John A., Ina V. S. Mullis, and Chancey O. Jones. *Can Students Do Mathematical Problem Solving? Results from Constructed-Response Questions in NAEP's 1992 Mathematics Assessment.* Washington, D.C.: National Center for Education Statistics, 1993.

Lambdin, Diana V. "The NCTM's 1989 Evaluation Standards: Recycled Ideas Whose Time Has Come?" In *Assessment in the Mathematics Classroom,* 1993 Yearbook of the National Council of Teachers of Mathematics, edited by Norman L. Webb, pp. 7–16. Reston, Va.: National Council of Teachers of Mathematics, 1993.

Mumme, Judy. *Portfolio Assessment in Mathematics.* Santa Barbara, Calif.: California Mathematics Project, University of California, Santa Barbara, 1990.

National Council of Teachers of Mathematics. *Assessment Standards for School Mathematics.* Reston, Va.: National Council of Teachers of Mathematics, 1995.

————. *Curriculum and Evaluation Standards for School Mathematics.* Reston, Va.: National Council of Teachers of Mathematics, 1989.

————. *Professional Standards for Teaching Mathematics.* Reston, Va.: National Council of Teachers of Mathematics, 1991.

Pandey, Tej. *A Sampler of Mathematics Assessment.* Sacramento, Calif.: California Department of Education, 1991.

Pennsylvania Department of Education. *Mathematics Assessment Handbook.* Harrisburg, Pa.: Pennsylvania Department of Education, 1993.

Seeley, Marcia M. "The Mismatch between Assessment and Grading." *Educational Leadership* 52 (October 1994): 4–6.

Stenmark, Jean Kerr, ed. *Mathematics Assessment: Myths, Models, Good Questions, and Practical Suggestions.* Reston, Va.: National Council of Teachers of Mathematics, 1991.

Vermont Department of Education. *Looking beyond "The Answer": The Report of Vermont's Mathematics Portfolio Assessment Program.* Montpelier, Vt.: Vermont Department of Education, 1991.

Webb, Norman L., ed . *Assessment in the Mathematics Classroom,* 1993 Yearbook of the National Council of Teachers of Mathematics. Reston, Va.: National Council of Teachers of Mathematics, 1993.

24

Mathematics Pen-Pal Letter Writing

Eileen Phillips

COMMUNICATION, in all its aspects and in all subjects, has become a major focus in my teaching over the last few years. One specific vehicle that I use to promote written communication in mathematics is mathematics pen–pal letters.

I teach a fourth-grade class (i.e., eight-, nine-, and ten-year-olds) in a fairly affluent area of Vancouver, British Columbia. My school is located close to the University of British Columbia (U.B.C.), which allowed a relatively easy exchange of mathematics pen-pal letters between my students and preservice teachers who are enrolled in a mathematics methods course at U.B.C.

WHY PEN-PAL LETTERS?

Much research supports the pedagogical use of writing in mathematics classes, mainly journal entries (e.g., Borasi and Rose [1989]; Clarke, Waywood, and Stephens [1993]; Countryman [1992]; Stempien and Borasi [1985]; Waywood [1992]). Although I use journal writing, I have found that the road to meaningful entries is a long one. In September, many students are barely able to write a description of what they've done. It takes much experience with writing before they reach the stage of developing questions, writing about their thinking, and extending ideas explored in class.

Believing that all students are capable of writing reflectively and wanting all my students to experience the pleasure of writing in order to "see" what they are thinking, I decided to give my class another writing experience. I had read about Fennell (1991) asking his university students to write to elementary school students to help them learn how to diagnose learning problems. It was thus easy for me to envisage my students writing to university students for the purpose of exploring thinking about mathematics.

I believed that my students would receive three benefits: they would be (1) writing about their mathematics experiences, (2) reading responses about mathematics, and (3) creating mathematics. I wanted to see whether pen-pal letters could support the development of writing reflectively. Would pen-pal letter writing and reading promote my students' movement through the stages from primarily descriptive writing to personally engaged, reflective writing?

SETTING IT UP

I contacted a friend who was teaching a mathematics methods course for preservice teachers at U.B.C. She was very enthusiastic about my idea and offered to ask her students to volunteer to be pen pals. Since the letters would be done on her students' own time and would not be for course credit, we were pleased to find that twenty-seven of her students volunteered. Because these students were extremely busy and had heavy schedules, I believe their response shows the appeal of this project to adults learning about teaching. Since I had twenty-nine students in my class, the teaching assistant from the university course also became a pen pal and one preservice student had two pen pals.

I told my students that they would be writing to university students who were learning to be teachers and that their role was to be advisors. I explained to them that they would be testing activities that their pen pals would give them and that they would be responsible for reporting back to their pen pals how they found the activity. That is, they were to focus on whether the activity was fun, too hard, too easy, or just about right. Also, if they would change the activity in any way, they were to suggest how. My students were also expected to share their thoughts about what we were doing in our mathematics classes with their pen pals. Additionally, they were encouraged to make up challenging questions, problems, or activities for their pen pals—ones that would demonstrate things they liked to do in mathematics.

THE EXCHANGE OF WRITING

My fourth-grade students initiated the letters, and each one was randomly paired with a preservice teacher. Their first letters included personal information, a description of our mathematics class, and a self-evaluation of how good a mathematics student they thought they were. They discussed their favorite and least favorite mathematics topics. (See fig. 24.1.)

When they received their first replies, excitement was high. It was amazing how quickly some pairings established rapport. It was also interesting to see the activities that preservice teachers thought would be engaging. These ranged from traditional computation questions and drills to graphing questions that required involving the pen pal's family or problems that involved toothpicks, pattern blocks, and other materials.

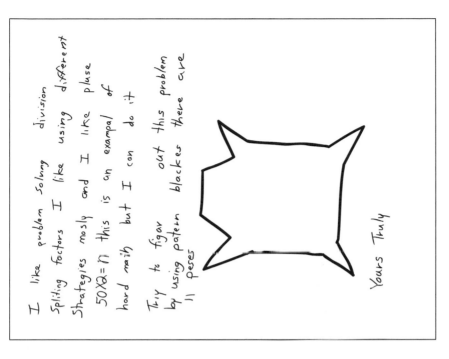

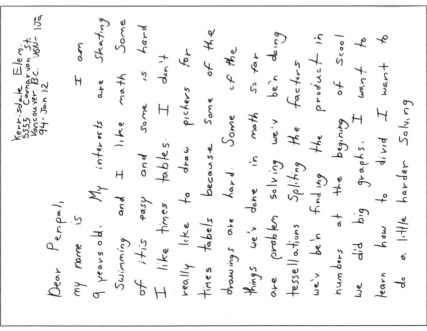

Fig. 24.1. Introductory letter from fourth grader to preservice teacher

Some preservice teachers added interest by brightening the work with colored felts and drawings, whereas others sent photocopies of text problems with no attempt at personalizing.

My role in this project soon developed into that of an interpreter of handwriting, a resource person for mathematics activity suggestions, and a supplier of materials, such as paper, string, and toothpicks. Behind the scenes I was also the photocopier of letters. This was necessary and important, so that my students and I could have a reference file of their previous letters. I also delivered and received the letters. We completed an exchange every two weeks. By the end of the two-month project, my fourth-grade students had written four letters and two directed writings, and they had received four letters.

WHAT THEY WROTE ABOUT AND HOW THEY WROTE IT

Before each letter-writing session, we reviewed, as a class, the ideas that we had explored in mathematics since the last letter. I encouraged my students to share these ideas with their pen pals, but it was not mandatory. I also asked them to respond to their pen pals' questions and to write about the activities that their pen pals gave them. (Again, though this was encouraged, it was not monitored by me.) I expected the pen-pal relationship to provide its own motivation. What was not optional was writing. Pen-pal letter writing was a required part of my students' mathematics program, and I made certain that deadlines were met.

Because the students were given guidelines that they could choose to follow or not, their letters remained their own. I did not proofread them for spelling, mathematical accuracy, or content. I did check to see that they were addressed and dated correctly and were signed.

It was interesting to see what developed. In general I found that the fourth-grade students insisted on being the advisors and leaders of the exchange. If preservice teachers tried to introduce a new subject or concept, it was often completely ignored. But if they responded to a topic that the fourth graders had introduced, the topic would be extended. If the adult pen pal was clever at giving a new twist to a fourth grader's topic, this strategy was also successful in promoting writing.

For example, one fourth-grade student wrote about some graphing that we had been doing, and her pen pal responded by supplying the activity in figure 24.2. In turn, the child then developed her own graphing problem, and so did one of her friends with whom she had shared her letter. I often noticed this ripple effect, and students effectively became teachers of their peers.

Also of interest was the method of responding. The preservice teacher's writing style and tone were generally taken as the model. If the adult pen pal was thorough at explaining and developing ideas, my student became much better at this, too. If the adult pen pal was reflective, my student mirrored this behavior. Whether the preservice teacher's writing was formal or relaxed, my student tended to respond similarly.

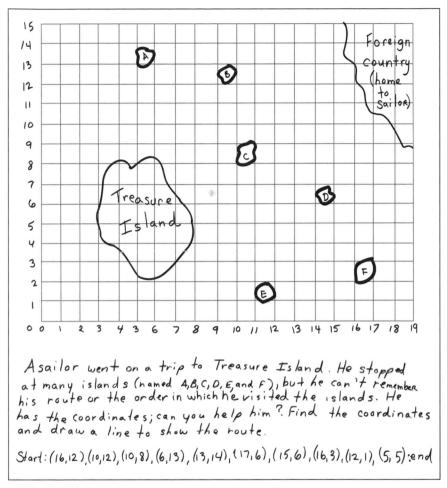

A sailor went on a trip to Treasure Island. He stopped at many islands (named A,B,C,D,E, and F), but he can't remember his route or the order in which he visited the islands. He has the coordinates; can you help him? Find the coordinates and draw a line to show the route.

Start: (16,12),(10,12),(10,8),(6,13),(13,14),(17,6),(15,6),(16,3),(12,1),(5,5):end

Fig. 24.2. A preservice teacher's activity written to a fourth-grade pen pal after hearing that graphing was being studied

This response seems to be the key to my initial research interest: Can pen-pal letters be used to support the development of reflective writing? It seems, on the basis of this project, that pen-pal letters can be used to promote reflective writing *if* the adult pen pal provides a reflective model. Reading reflective writing, combined with a genuine need to establish a dialogue in writing, seems to be a primary element in developing reflexivity.

Twice I asked my students to describe and explain specific topics to their pen pals. These topics were included in the project as "directed writings" rather than as letters and were called "Lattice Multiplication" and "My Opinions about Calculators." I wanted to see how my students would deal with these ideas. In a previous year I had asked students to write about these subjects in their mathematics journals and had met with reluctance—reluctance not to sharing but to writing such a lengthy amount. This time, however, the students happily drew diagrams and wrote explanations, in prose or even poetry, to represent their opinions and understandings.

A PEN-PAL PARTY

Just before the mathematics methods course ended for the preservice teachers, we invited our pen pals to visit us. Mathematics activities were prepared, food was planned, and welcome signs were made. In the evaluation survey given at the end of the project, many students in both groups commented on how great it was to meet their pen pal (fig. 24.3).

Most pen-pal pairs used their time to explore and extend problems that had been developing over the course of their writing. Often the face-to-face meeting was a way of continuing dialogue that had started through writing. I found that extra activities that had not been developed or introduced in the letters were not needed.

Fig. 24.3. Our pen-pal party

CONCLUDING COMMENTS

This was an extremely rich project. For me, it furnished insights into my students' attitudes and an in-depth look at their thinking. I was reminded, once again, of how capable students can be at creating and adapting ideas for their own learning. My belief that pen-pal letters can be an effective vehicle for promoting reflective writing was justified.

For my students, the project provided a reason for writing about mathematics, a chance to read writing about mathematics, an opportunity to be really listened to as an advisor, and a look at the world of a university student. In *Speaking Mathematically*, David Pimm (1987) draws attention to some of the benefits of writing thoughts down. For many of my students this was the first time that these benefits made sense.

> [W]riting serves as a record/reminder of thought…. Writing also externalizes thinking even more than speech by demanding a more accurate expression of ideas. By writing something down, it then becomes outside oneself and can more easily be looked at and reflected upon, with all the benefits of a visible permanent record. (P. 115)

For the preservice teacher, this project created a window into one mathematics classroom and an opportunity to explore mathematics with one student. As Brown (1981) states, "One incident with one child, seen in all its richness, frequently has more to convey to us than a thousand replications of an experiment…" (p. 11). Furthermore, it also often gave the preservice teacher the opportunity to write about mathematics for the first time in her or his life. It also gave them, as a group, the opportunity to see the kinds of things students at this level were interested in and to view the whole range of writing and mathematical abilities spread before them.

REFERENCES

Borasi, Raffaella, and Barbara J. Rose. "Journal Writing and Mathematics Instruction." *Educational Studies in Mathematics* 20 (November 1989): 347–65.

Brown, Stephen I. "Sharon's Kye." *Mathematics Teaching* 94 (March 1981): 11–17.

Clarke, David J., Andrew Waywood, and Max Stephens. "Probing the Structure of Mathematical Writing." *Educational Studies in Mathematics* 25(3) (October 1993): 235–50.

Countryman, Joan. *Writing to Learn Mathematics.* Portsmouth, N.H.: Heinemann, 1992.

Fennell, Francis. "Diagnostic Teaching, Writing and Mathematics." *Focus on Learning Problems in Mathematics* 13 (Summer 1991): 39–50.

Pimm, David J. *Speaking Mathematically.* London: Routledge, 1987.

Stempien, Margaret, and Raffaella Borasi. "Students' Writing in Mathematics: Some Ideas and Experiences." *For the Learning of Mathematics* 5 (November 1985): 14–17.

Waywood, Andrew. "Journal Writing and Learning Mathematics." *For the Learning of Mathematics* 12 (June 1992): 34–43.

25

Developing Preservice Teachers' Strategies for Communicating in and about Mathematics

Frances R. Curcio

Sydney L. Schwartz

Catherine A. Brown

MEANINGFUL communication in mathematics is characterized by discourse that promotes inquiry into an idea. Such inquiry involves discovering underlying relationships between and among mathematics concepts and takes the form of thinking and discussing, manifested in oral and written communication. Genuine mathematics learning entails constructing meaning by making connections between prior understandings and new experiences (NCTM 1991). Discourse brings individuals' intuitive understandings to the conscious level, exposing ideas for examination and discussion with others and creating an opportunity for knowledge construction (Dunn and Larson 1990). Affording learners opportunities to share their mathematical ideas as they work with diagrams, pictures, graphs, symbols, concrete models, and analogies supports the construction of meanings for each learner (Kamii 1985). From this perspective, effective instruction in mathematics involves dynamic discourse among teachers and learners.

Although experienced mathematics teachers may be comfortable in an environment characterized by student interactions, preservice teachers often find integrating communication in and about mathematics a challenge. Typically, in teacher education, two types of experiences provide opportunities for preservice teachers to engage in learning about discourse in teaching mathematics—methods courses and field work. After experiencing more than 14 000 hours as students observing teaching in elementary and secondary school classrooms (Ciscell 1994), prospective teachers begin a methods course with a defined set of perceptions about the teaching and learning of mathematics, including the role of discourse. The experiences they bring are influenced by both accurate and inaccurate conceptions, many of which are not at the conscious level. The methods instructor is faced with the task of challenging these conceptions about mathematics and discourse by offering experiences that expose concepts for examination and, in the process, either strengthen or redirect them (Ball 1988).

In the ideal situation, field work, offered in conjunction with methods courses, includes peer teaching, observing class interactions, interviewing or tutoring children, microteaching, and student teaching. If preservice teachers are to develop the necessary skills, discourse experiences need to be integral to all field-based activities as a way of using and extending the knowledge and skills developed in the methods course. Field work and methods courses that are grounded in a common philosophy can offer a unified and consistent message about learning and teaching mathematics, an essential ingredient for effective teacher education (Eisenhart et al. 1993).

This article presents a coordinated series of methods-based and field-based activities designed to help preservice teachers develop understandings and skills that include discourse as an integral strategy in mathematics instruction. Discussed below is the rationale for developing three kinds of activities in which preservice teachers need to engage as they shift their primary role from that of learner to that of teacher. The activities comprise three steps in a process for developing power in using discourse to teach mathematics: (1) experience discourse as learners, (2) reflect on discourse, and (3) use understandings about discourse to design inquiry-oriented teaching. This final step requires the preservice teachers to transform their knowledge of discourse into instructional plans. Although reflection is increasingly included in preservice teacher education, this focus on reflection rarely includes discourse.

Learning: Experiencing Discourse

Mathematics methods instructors use several approaches to help preservice teachers learn to teach mathematics. In one technique, the instructor sets the task, supplies the concrete materials, and describes how to use the materials in fulfilling the task. Working in pairs or in small groups, preservice teachers manipulate the materials and together engage in a process of examining important mathematical ideas. Jointly, they construct and interpret meanings about embedded mathematical relationships. Spontaneous discourse between and among preservice teachers occurs during the process of completing the mathematical task, and the instructor participates periodically in some of this discussion. The instructor then leads a more focused, follow-up discussion, asking participants to share how they approached the task and to compare their results. The dual function of this approach is to foster mathematical understandings while simultaneously modeling methods for teaching mathematics using discourse and concrete materials. This approach flows from the belief that the best way to help future teachers develop strategies for dialogue with students is to engage them in experiences that will convince them to employ the model when they make the transition from the role of learner to that of teacher.

Another way for instructors to engage preservice teachers as learners of discourse is to give them various materials to examine mathematically. Depending on the concepts to be developed, the materials may be buttons, blocks, interlocking cubes, Styrofoam packaging pieces, boxes and jars of different sizes, or wallpaper samples. As the preservice teachers work in pairs, the instructor asks them to describe their responses to the following question: "How do these materials 'speak to you' in mathematics?" Figure 25.1 contains a list of responses produced by preservice teachers during different class sessions.

As the preservice teachers examine the concrete materials, the instructor promotes understanding by—

- observing preservice teachers as they talk about ideas and by extending the discussion with a provocative question. For example, observing the cubes in uniform groups of ten, the instructor may ask, "Do all the colors have to be the same?"
- encouraging students to summarize and validate their thinking. For example, sitting down with one group, the instructor may ask, "What mathematics concepts have you discovered as you worked with these materials?" or "Do you all agree that these mathematical ideas are represented in these materials?"
- redirecting a search for an answer. For example, after being bombarded with preservice teachers' requests for a definitive

Concrete Materials	Mathematics Content Represented
buttons	numerosity; multiples; attributes (e.g., size, color, shape, number of holes) for classifying and sorting; geometric shapes
blocks	attributes for classifying and sorting; geometric shapes
interlocking cubes	numerosity; nonstandard measurement units; place value
Styrofoam packaging pieces	numerosity
boxes and jars of different sizes	surface area; volume
wallpaper samples	geometric shapes; tessellations; area; geometric transformations (e.g., slides, flips, turns)

Fig. 25.1. Preservice teachers' responses to the question, "How do these materials 'speak to you' in mathematics?"

answer about the unit being transformed to generate a wallpaper pattern, the instructor might ask, "Is there only one unit that could generate this pattern?"

After the pairs share their ideas with the rest of the class, they form small groups of four or five and work on mathematical tasks, selecting the materials they wish to use. Following the investigation, the groups share their results and describe reasons for selecting the materials. Figures 25.2 and 25.3 contain preservice teachers' responses to some mathematical tasks given in an elementary grades mathematics methods course and in a middle grades mathematics methods course, respectively.

Mathematical Task

Figure out as many different ways as you can to solve the following word problem and record your results.

Jeffrey has 6 pieces of bubble gum. How many more pieces of bubble gum does he need to have 15 altogether?

Debbie: "We counted out 6 Styrofoam pieces, keeping them together in one pile. Then we continued counting from 7 to 15, keeping the remaining 9 in a separate pile. Any counters could have been used to represent the bubble gum. We decided to use Styrofoam pieces because we pretended they were pieces of gum."

Juliet: "I started at 6 and then raised 9 fingers as I continued to count to 15. I find it convenient to use my fingers."

Maureen: "I used the symbols, $6 + \_\_\_ = 15$, so I knew there were 9 pieces of gum needed. I always try to translate a word problem into an equation."

Fig. 25.2. Different ways three elementary grades preservice teachers solved a word problem

In leading the discussion for the whole group, the instructor may focus on having the students share the variety of strategies they employed to fulfill the requirements of the task, asking, for example, "Does anyone have another strategy?" After sharing their strategies, the instructor may ask them to compare, "Are some of these strategies really the same or are they different?" "How are they the same?" "How are they different?" In sustaining the discourse, the instructor may engage preservice teachers in making connections between the alternative strategies and the different materials, asking, for example, "How many different ways can you demonstrate this concept with the materials?" The content of the posttask discussion includes important mathematical ideas and relationships that are accessed by using the typical materials listed in figure 25.1. This particular approach distinguishes itself from the previous approach by involving the preservice teachers in constructing their own mathematical meanings from the use of the concrete materials, without having the meanings imposed on them by the instructor.

Mathematical Task

In your groups, find as many different methods as you can to determine the volume of Box A (a 10 cm × 12 cm × 6 cm open-top box). Record your methods and be ready to share them with the entire class.

Response written by group 3:

Method 1: "We remembered that volume equals length times width times height, so we measured the box and got 10, 12, and 6 centimeters and then multiplied 120 times 6 and got 720. And since it's volume, we made it 720 centimeters cubed."

Method 2: "We thought about covering the bottom of the box with a layer of centimeter cubes and then seeing how many layers would fit in the box by stacking cubes up in one corner. We found that one layer was 120 cubes and that that layer and 5 cubes would stack up. So we figured 6 layers, each with 120 cubes. 6 × 120 is 720, 720 centimeter cubes."

Method 3: "We did something like method 2 except we covered one side of the box with cubes. It took 60 cubes. It took 60 cubes to cover the side, and then we laid down cubes to see how many we could fit and it was 12. Then we knew that it would take 60 times 12, or 720, cubes, so the volume is 720 cubic centimeters."

Method 4: "We did the same thing as method 2 and 3 except we covered the other side and found it took 72 cubes. There would be 10 layers of 72 and that gave us 720 cubic centimeters again!"

Fig. 25.3. Different ways a group of middle grades preservice teachers determined the volume of a box

REFLECTING: DEVELOPING UNDERSTANDING THROUGH DISCOURSE ABOUT DISCOURSE

In the two approaches described above, both instructors orchestrate two different types of discourse. First, they set up a context for preservice teachers to talk among themselves to discover mathematical relationships or ideas, and jointly, to make meaning. Second, the instructors lead a group discussion focusing on what preservice teachers learned from the mathematical tasks. Both these discourse formats expose preservice teachers to pedagogy involving concrete materials, but neither provides them the opportunity to reflect on the different ways in which the instructors orchestrate or encourage discourse as a pedagogical strategy.

Preservice teachers can build a knowledge base of discourse strategies in many different ways. Reflecting on the impact of discourse on their own experiences as learners sets an ideal context to identify how discourse contributed to their own learning. In the illustrations about teaching in mathematics methods courses described above, students engaged in discourse during the period when they were collaborating and conversing spontaneously about the task and again when the total group shared their knowledge and ways of knowing.

Engaging in the process of identifying discourse strategies used in instruction focuses attention beyond the mathematical content to the pedagogical role in a learning encounter. One script for building experiences of reflection on discourse might be as follows:

> After preservice teachers complete a discussion of their own learning and an analysis of the opportunities for teaching mathematics using concrete materials, the instructor directs attention to the conversations that were initiated during the period of working on the task using materials. Working together again, preservice teachers recall and list comments and questions initiated by the instructor that provoked further thinking and discussion.

When we provoke preservice teachers to reflect on an instructor's discourse strategies, they identify pedagogical approaches that support—

- observing and extending. For example, the preservice teachers stated, "You came over, sat down and asked us what we found out about the mathematics we thought we could represent in the wallpaper. After we told you about the unit we found that could be transformed to create the pattern on paper, you asked if that was the only mathematical concept we could find in the wallpaper sample. That got us to brainstorming about other concepts—and there were a lot of them!"

- summarizing and validating. For example, the preservice teachers commented, "When you came over, you listened and then you said, 'Sounds like you're having a really good discussion here.' Then you jumped in with a question, 'Is that the only way you can do it?' "

- redirecting. For example, the preservice teachers remarked, "Every time we asked you a question, you answered with another question. This was driving us crazy. We know we have to think of the answers for ourselves."

Once this list has been generated, the instructor can change the focus, asking the preservice teachers to recall and list the strategies the instructor used with them to initiate and extend discourse during the total group discussion period. Finally, the preservice teachers compare the strategies on the two lists, grouping and clustering similar ones and distinguishing those used in the spontaneous discussion from the strategies associated with the more formalized total group discourse.

It is possible to build on these reflections on discourse, to examine through reflection how individuals vary in their reactions to different strategies. For example, when asked, "How did you figure that out?" some preservice teachers report that they find such a question anxiety producing, whereas others find it challenging and stimulating. Further reflection by preservice teachers leads them to generate lists of alternative ways to ask the same question, such as, "Would you be willing to share your thinking with us about how you went about solving the problem?"

or, "What steps led you to that conclusion?" or, "When you were solving the problem, what did you do first?"

Additional experiences that help preservice teachers become attuned to the strategies of discourse designed to provoke thinking and discussing in mathematics can include viewing videotapes of teachers engaging in discourse with class groups at different grade levels, featuring different mathematics topics. Seeing others use discourse after reflecting on one's own experiences as a learner creates an opportunity to become more skillful in recognizing strategies of discourse. Reflecting about discourse encourages preservice teachers to develop their understanding of the role of discourse in learning and teaching.

TRANSFORMING DISCOURSE EXPERIENCES: FROM LEARNER TO TEACHER

Generating understandings through reflection is a prerequisite to transforming ideas about discourse into use. Reflection does not automatically lead to a transformation of understandings from the perspective of the learner to the perspective of the teacher. To accomplish this transformation, it is necessary to offer preservice teachers opportunities to plan for, implement, and review their own teaching experiences designed to promote discourse.

One way to offer transformational experiences is to have preservice teachers working in pairs write out discourse possibilities related to launching a mathematical task. In the process of jointly planning, the prospective teachers are asked to generate possibilities for initiating and sustaining discourse. To do this, they need to focus on concepts appropriate for a particular grade level and establish a context suitable for engaging children in a meaningful task. Examples of elementary and middle grades activities designed and implemented by preservice teachers follow.

Example of an Elementary Grades Mathematics Activity

Two elementary preservice teachers, Ken and Maureen, designed and implemented the following plan:

Grade level: Third grade

Context: Setting up a store for children to buy holiday gifts for family and friends

Mathematics content: Estimation; monetary equivalences and exchanges; joining and separating amounts of money

Materials: Play money; items for sale (dolls, $2.00; toy cars, $1.50; Frisbees, $3.00; balls, $1.00; baseball caps, $5.00; blowing bubbles, $0.50; balloons, $0.25)

Task: Given $10.00, make a list of friends and children in your family for whom you plan to buy presents. Decide how much you want to spend for each person. Buy your gifts.

Questions for discussion: "Are you going to spend the same amount of money for each person on your list or will some people be just as happy with less expensive gifts?"

Discourse extenders: As children work in pairs, listen to the problems the children identify (e.g., not enough money to buy the gifts they want). "Are you both buying the same number of presents or a different number of presents? Do you plan to spend the same amount of money on each person?" For children who do not have enough money, ask, "What changes will you have to make so that you do have enough money to buy presents for everyone on your list?"

Summary: In a whole-group setting, ask, "Let's find out how many gifts everybody bought?" "Who bought the most gifts?" "Who spent the most money?" "Who bought the fewest number of gifts?" "Who spent the least amount of money?" "Who bought the most expensive gift?" "Who spent a different amount of money for each gift?" "What happens to your money when you buy the most expensive gifts?" "Did you notice what happens when you change the amount of money you spend for each gift or when you change the number of gifts you need to buy?"

As Ken conducted the activity, Maureen observed and kept a record of the interactions, noting which strategies kept the children talking and thinking about mathematical ideas and which ones did not. The challenge for Ken emerged when some children struggled with a variety of problems. Ken was tempted to offer solutions rather than have the children work out the problems themselves. In some cases, he asked the same questions repeatedly and was unable to make any progress with the children. Maureen and Ken discussed what could have been done. Although Maureen did not intervene, some observers or recorders do jump in to assist in relieving such a "log jam." If videotape equipment had been available, Maureen could have documented the interactions on videotape, which could later serve as a vehicle for reflection and discussion.

Example of a Middle Grades Mathematics Activity

Two preservice teachers, Shanda and Mark, designed and implemented the following plan:

Grade level: Seventh grade

Time: Two to three class periods

Context: The school parents' group has decided to ask students to sell wrapping paper to make money to buy new calculators for the mathematics classes. They have found a company that is willing to include three designs by students in the selection of wrapping paper they make for the school to sell. The mathematics teachers and art teachers have been asked to help pairs of students create designs that will be entered in a contest to be judged by two local interior designers. The three winning pairs will have their designs made into wrapping paper to sell.

Mathematics content: Geometric shapes, symmetry, transformations

Materials: Large sheets of newsprint, samples of wrapping paper and wallpaper, geometric shapes cut out of tag board, rulers, compasses, protractors or angle rulers, colored pens, pencils, and crayons

Task: Using the materials available, create a design for "all occasion" wrapping paper. The design should be such that it could be used for wrapping paper that comes on a roll as well as in sheet form. After each pair of students has at least one design, we will critique each design as a class. As a class, we will construct some guidelines for designing wrapping paper, and each pair will create another design.

Questions for discussion: "Is there anything you would think about when creating a design for paper on a roll that is different from what you may consider when designing sheets of wrapping paper?" "What characteristics make wrapping paper attractive or interesting to you?"

Discourse extenders: As students work in pairs to design paper, listen to the discussions the students have about what patterns are best to make and how to construct them. Ask questions like, "Do you both agree on what characteristics wrapping paper on rolls must have?" "How could you try out a design for a roll of paper?"

Summary: In a whole-group setting, after most pairs have created at least one design, each pair will be asked to present their design and the class will discuss the designs. The teacher can instruct each pair, "Please describe your thinking as you created your design and what you think makes it attractive or interesting." If the students have difficulty presenting their design, ask, "Is there a pattern in your design that repeats?" "How did you come up with the pattern?" "How does your use of color contribute to your design?" "Who do you think will like your design? Young children? Teenagers? Adults?" "What makes you think so?" Following each presentation, ask the class, "What characteristics of this design do you find attractive or interesting?" and write the responses on the board to be used later in determining the class guidelines for designing wrapping paper.

Shanda conducted the activity in the grade 7 class in which she was a student teacher that semester. It took three days to complete the activity. Mark observed and videotaped the lesson during the three days. After each session, Mark and Shanda talked about how the pairs of students worked and how the class presentations and discussions were going. After the second session, in which Shanda had a difficult time getting students to listen to presentations and to participate in the class discussion, Mark and Shanda watched the videotape together and brainstormed about how Shanda might manage that activity better during the third session. After the third session, Shanda and Mark wrote a joint reflection on the experience.

In this third type of activity, preservice teachers are expected to transform their understandings into plans for classroom activities. They are asked to transform their knowledge of mathematics, materials, and discourse, gained through their experiences as learners, into plans for activities they can implement as teachers in a classroom. What we are looking for is fluency in developing discourse strategies so that preservice teachers may continue to grow. This is an important step in the process of becoming a teacher of mathematics and one that requires assistance from instructors. Ideally, as in the interactions between Maureen and Ken and between Shanda and Mark, preservice teachers' plans are implemented in a field placement, and the implementation is analyzed, reflected on, and revised, completing a cycle of activities that could be repeated a number of times.

CONCLUDING COMMENTS

The ideas discussed in this article suggest ways to help preservice teachers experience discourse as learners, reflect on discourse to develop understandings, and transform their knowledge of mathematics, materials, and the role of discourse into skills for instructional activities. As noted earlier, many teacher educators are very familiar with the general strategies of discourse presented but often use them exclusively as the strategies relate to teaching about mathematics instruction. It is often assumed that having experienced it, preservice teachers will be able to model it. The approach featured here highlights ways to develop useful strategies for helping preservice teachers gain understandings of, and skills in, discourse as a critical component of mathematics teaching and learning.

REFERENCES

Ball, Deborah. "Unlearning to Teach Mathematics." *For the Learning of Mathematics* 8 (February 1988): 40–48.

Ciscell, Robert. "Preparing to Teach: A Little Knowledge Is a Dangerous Thing." *Kappa Delta Pi Record* 30 (Winter 1994): 55–57.

Dunn, Susan, and Rob Larson. *Design Technology: Children's Engineering.* New York: The Falmer Press, 1990.

Eisenhart, Margaret, Hilda Borko, Robert Underhill, Catherine Brown, Doug Jones, and Patricia Agard. "Conceptual Knowledge Falls through the Cracks: Complexities of Learning to Teach Mathematics for Understanding." *Journal for Research in Mathematics Education* 24 (January 1993): 8–40.

Kamii, Constance Kazuko. *Young Children Reinvent Arithmetic: Implications of Piaget's Theory.* New York: Teachers College Press, 1985.

National Council of Teachers of Mathematics. *Professional Standards for Teaching Mathematics.* Reston, Va.: National Council of Teachers of Mathematics, 1991.

26

Strategies to Support the Learning of the Language of Mathematics

Rheta N. Rubenstein

Mathematics uses many terms. How can we make learning these terms more interesting for students and more effective in building understanding? Some strategies I have found successful include inviting students to invent their own terms, exploiting analogies and metaphors, using charts, and revealing the origins of language.

Inviting Students to Invent Their Own Terms

Students' invented language can crystallize ideas for them and promote understanding. For example, in a high school geometry class, students had learned that a midpoint is a point that divides a segment into two congruent parts. Later they learned that an angle bisector is a ray that divides an angle into two congruent angles. One student, Cecile, asked, "Instead of 'angle bisector,' why not say, 'midray'?" The class liked the idea. They replaced five syllables with two, highlighted the fact that they were talking about a ray, and recognized the analogy with a midpoint.

Later in the year, in a unit on loci, students were able to supply the correct terms for "the locus of points equidistant from two parallel lines in a plane" and "the locus of points equidistant from two parallel planes in space." Without hesitation, they answered *midline* and *midplane,* words never before uttered in the room. I was stunned. Their simple, descriptive phrases replaced my long-winded descriptions and communicated perfectly. I discovered that the earlier invention of *midray* had helped the students internalize the bisection notion at a deep and meaningful level.

In another instance, the students were studying major and minor arcs. One student asked, "If we call angles that total 90 degrees *complementary* and angles that total 180 degrees *supplementary,* why isn't there a name for arcs that total 360 degrees?" I was delighted by the question but had no inkling of an answer. I invited the class to think about what term would be good. A few days later, Byron suggested *circlementary.* The class loved it. It agreed in structure with the related terms and, even better, clearly identified the notion of the completeness of the circle. Years later I learned a formal term for such pairs, *explementary.* What an opportunity I would have lost if I had just told it to the students!

One concern sometimes raised about invented language is that students will be unfamiliar with the conventional terminology. To avoid this problem, I ask students to think about why we need conventional terms. They note that without agreement we'd have a Tower of Babel—the inability to communicate. So students appreciate that although their imagination and input are valued, we have to "translate" occasionally into conventional terms to be sure we can communicate with the rest of the world.

Invented language has many benefits. The major one is that it can enhance understanding. In the examples cited, the students were engaged in making sense of mathematical ideas and found that doing it "in their own terms" made ideas clearer and easier to grasp. Another benefit is student ownership. Students realize that they are links in a chain of mathematical thinking and that they can participate in the invention and development of mathematics.

EXPLOITING ANALOGIES AND METAPHORS

Analogies can be another helpful tool for assisting students in making sense of new terms and ideas. Completing exercises like the following at appropriate points in students' studies builds connections between known and new ideas and invites higher-level thinking.

collinear : line :: _____ : plane
_____ : 1D :: coplanar : 2D
congruent : _____ :: equal : numbers
phrase : sentence :: expression : _____
sphere : hemisphere :: circle : _____
prism : cylinder :: pyramid : _____
plane : circle :: _____ : sphere

(*Solutions:* coplanar, collinear, geometric figures, equation, semicircle, cone, space)

Other analogies can be discussed informally. For example, when students are first introduced to using a coordinate system, some need help remembering to go across first, then up or down. Noting the analogy with reading a book often helps: first read across the page, then down. Analogies also help in teaching transformations. Students are often puzzled that $y = (x - h)^2$ is a translation of $y = x^2$ by h units to the right and not to the left. In addition to creating tables and graphs, some students benefit from making the analogy with daylight savings time that when you move your clock back, you gain an hour.

USING CHARTS

Frequently, a well-chosen chart assists students in seeing mathematical relationships. For example, students can be asked to generate a 3 × 3 matrix (see fig. 26.1) to help them visualize the categories of triangles and recognize that triangles can be classified in two distinct ways—by sides and by angles. The fact that some intersections on the matrix are impossible (i.e., the cells are empty) challenges their thinking, too.

Another chart that has been powerful in unifying a number of relationships and sorting out the language of quadrilaterals is the hierarchy in figure 26.2 (Craine and Rubenstein 1993). (A middle school version of this activity can be found in Geddes et al. 1992.) The shapes above are special cases of those below. (Note that the hierarchy uses an inclusive definition of *trapezoid* as a quadrilateral with *at least* one pair of parallel sides.) For example, a rhombus is both a kite and a parallelogram. Properties are inherited by the shapes that are higher on the chart. An isosceles trapezoid can be defined as a trapezoid with congruent base angles. This property is passed up the chart to the rectangle and square.

As students progress in their studies, the hierarchy organizes a growing number of ideas, such as the properties of symmetry and diagonals. It, and other visuals that are developed during a course, make good bulletin boards that evolve as new ideas and terms are incorporated.

REVEALING THE ORIGINS OF LANGUAGE

The origins of words are often helpful in bringing language to life, making terms more meaningful, and revealing connections with related ideas. *The Words of Mathematics* (Schwartzman 1994) is an excellent resource that was used in verifying the following examples:

- I ask students who confuse *horizontal* and *vertical* to think of the position of the horizon.
- In trigonometry, exercises that reveal that *cosine* comes from "complement's sine" link what is usually taken as a property to the meaning of the term.
- Researching the origin of the word *average* (from the Arabic *awariyah,* "goods damaged in shipping,") leads to the story that when shipped goods didn't reach port, investors shared the losses proportionally, thus generating the word for a total divided by the number of parts composing it.

By Sides / By Angles	Equilateral (Three Congruent Sides)	Isosceles (Two Congruent Sides)	Scalene (No Congruent Sides)
Right			
Obtuse			
Acute			

Fig. 26.1. Classifications of triangles

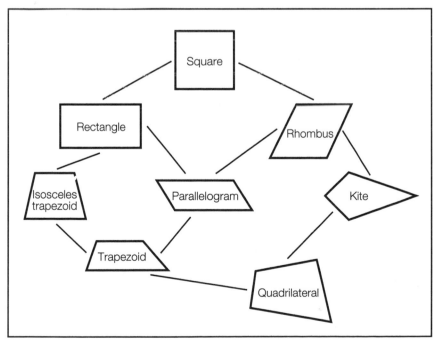

Fig. 26.2. Quadrilateral hierarchy. Reprinted with permission from Craine and Rubenstein (1993)

• Noting the similarity between a trapezoid and a trapeze—the circus acrobat's perch that has a beam parallel to its crossbar—helps students focus on the parallel sides.

Other terms share a root, a prefix, or a suffix. Listing words with related parts helps build connections and reduce the number of things to learn.

• *Bi-*, which means "two," occurs in *binary* (base-two numeration), *bisect* (cut into two equal parts), *binomial* (the sum of two terms), and *bifurcation* (branching into two categories).

• *Iso-*, which means "same," occurs in *isosceles* (same sides), *isomorphism* (same form or structure), and *isometric* (same measure).

• *Dia-*, which means "across," occurs in *diameter* (a measure across a circle) and *diagonal* (a segment that goes across from one angle or vertex to another).

My favorite word origin, however, is that of the word *mathematics*. Although many students know that *geometry* means "earth measure," some know that *trigonometry* means "triangle measure," and a few know that *algebra* comes from the Arabic for "the reunion of broken parts," very few students are aware of the etymology of the name of our discipline. *Mathematics* means exactly what educators care strongly about: learning!

CONCLUDING COMMENTS

Learning the terms of mathematics need not be a burden. Words are a natural part of human activity; they have histories, relations to one another, and connections to the real world. Students can appreciate language and value its role in supporting communication and understanding when they are engaged in inventing, visualizing, and studying the history, uses, and connections of words.

REFERENCES

Craine, Timothy, and Rheta Rubenstein. "A Quadrilateral Hierarchy to Facilitate Learning in Geometry." *Mathematics Teacher* 86 (January 1993): 30–36.

Geddes, Dorothy, with Juliana Bove, Irene Fortunato, David J. Fuys, Jessica Morgenstern, and Rosamond Welchman-Tischler. *Geometry in the Middle Grades. Curriculum and Evaluation Standards for School Mathematics* Addenda Series, Grades 5–8. Reston, Va.: National Council of Teachers of Mathematics, 1992.

Schwartzman, Steven. *The Words of Mathematics: An Etymological Dictionary of Mathematical Terms Used in English.* Washington, D.C.: Mathematical Association of America, 1994.

27

Communication in Mathematics for Students with Limited English Proficiency

Rafael A. Olivares

During the last decade, the number of children who speak other languages who had difficulty speaking English in U.S. schools increased 27 percent, from 1.9 million to 2.4 million (U.S. Department of Education, National Center for Educational Statistics 1994). To succeed in mathematics, these students have to participate actively in mathematics classes in spite of having limited English skills (Santiago and Spanos 1993). Educators' experience shows that their English proficiency in day-to-day interactions and their ability to translate the mathematics lexicon from the first language (L1) into the second one (L2) is not sufficient to meet the requirements of the mathematics classroom (Chamot and O'Malley 1993; Cuevas 1991). What these students need is to acquire the language of mathematics in English. This paper discusses the problems faced by students with limited English proficiency (LEP students) when they are learning to communicate mathematically in L2. To help teachers understand those problems, the paper describes the conditions in which LEP students develop communicative competence in mathematics. It also suggests a model that links communicative competences with L1, L2, and the prior mathematical knowledge and behaviors required to learn mathematics. On the basis of that model, several recommendations are made for classroom practice.

For nonnative speakers of the language, three characteristics make communication in mathematics different from everyday communication in English. First, to communicate mathematically, LEP students are required to work with abstractions and symbols. The representation of those abstractions and symbols does not always facilitate comprehension in day-to-day speech; consequently, LEP students cannot use the same clues they use in everyday interactions. For example, in day-to-day interactions, LEP learners can infer meaning from body language, the environment, and the context of the message. In those situations, they can use their intuitive notions about the world to infer meaning. That is not so with mathematics communication. Mathematical language and symbols are abstract and high in concept density (Fox 1977). Second, most of the time, each element of a mathematics proposition is fundamental for understanding the whole proposition. In day-to-day interactions, LEP students can guess and infer by understanding a few elements of the whole message. Understanding or making inferences about the whole without understanding each part in a mathematics proposition is difficult, if not impossible (Cuevas 1991). Third, the elements of mathematics propositions are frequently so specific that they cannot be rearranged. Rearrangements of the elements to comprehend the meaning of a message, as frequently happens in day-to-day speech, are scarce in mathematics. In this paper, a theoretical framework is offered that connects aspects of communicative competence in mathematics with the mathematical knowledge and behaviors of LEP students (see fig. 27.1).

COMMUNICATIVE COMPETENCE IN MATHEMATICS

Communicating mathematically in L2 requires from LEP students more than the ability to decode the English language (Ramirez and Chiodo 1994; Secada 1990). As stated in the *Curriculum and Evaluation Standards for School Mathematics* (National Council of Teachers of Mathematics [NCTM] 1989), to communicate mathematically is to think and communicate "about" mathematics. Communicative competences in the language of mathematics are at the core of the mathematical learning process and consequently are prerequisite to developing mathematical thinking. In other words, communicative competence in the language of mathematics is a necessary condition for mathematics learning. If the LEP student is not able to communicate "about" mathematics in L2, then there is no authentic mathematics learning in that language. According to Kessler (1987) four competences are necessary for mathematics communication: grammatical competence, discourse competence, sociolinguistic competence, and strategic competence. In short, these competences can be described as follows.

Grammatical Competence

To master grammatical competence, LEP students need to learn the lexicon and the structure of mathematics communication in English. The lexicon is the mathematics vocabulary, and the structure is the mathematics syntax used in English.

Vocabulary

Mastering the mathematics lexicon presents some peculiar learning problems for L2 learners. For example, specific terms of the mathematics lexicon can be transferred from one language to the other. For native Spanish

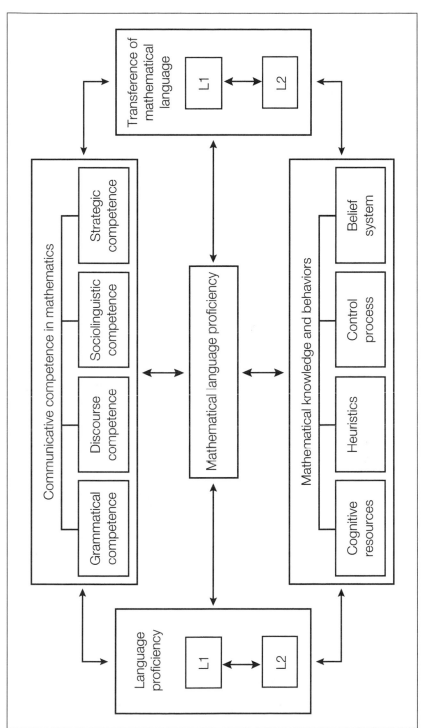

Fig. 27.1

speakers, "integers" are "enteros," "digits" are "dígitos," and "sets" are "conjuntos." The process requires changing the word representing the concept (label) from the L1 code to the L2 code. The concept remains the same; only its label changes. The meaning of addition (+), multiplication (×), and other symbols is the same in L1 and L2. In this instance, there is a positive transference between L1 and L2 (Geva and Ryan 1993). Transference, however, is complicated when LEP learners have to deal with words that are common in everyday English but have a different meaning in mathematics. For example, an LEP student who knows the meaning of the term *table* (furniture) in English has to add other meanings of that term when learning mathematics. The same happens with native speakers of English, but many times those two meanings are not represented by the same word in LEP students' L1. For example, in Spanish the equivalent for *table* (furniture) is *mesa*. The equivalent for the mathematical meaning of *table* is *tabla*. However, *tabla* in Spanish also means *board* and not furniture. Transference could work for the concept but not always for the terminology.

Another source of difficulties for LEP students is the use of synonyms in mathematics. For example, the concept of addition can be expressed in several different ways. LEP learners can be introduced to such words as *add, plus, combine, and, sum, increased by,* or *join.* Since all these words could represent the concept of addition, chances are that the nonnative English speakers do not know the equivalent of each of those terms in their L1. If they know them, the connection still has to be made between the L1 term and the mathematically correct equivalent in English.

Syntax

Students who master the mathematics lexicon in L1 are able to transfer to L2 a great deal of that vocabulary with a change of labels. However, learning the syntax of mathematics communication in English is a major problem for all LEP students (Tamamaki 1993). For example, reading in other languages may require following the text from top to bottom or from right to left. English text is read from left to right. Reading in mathematics occurs in different directions. Arithmetic algorithms and other forms of mathematics expressions could require reading right to left, top to bottom, bottom to top, diagonally, around, or following the arrow (Hoover and Nolan 1993).

Although there is a great deal of transfer of grammatical competence, some features of L1 can interfere with mathematics education in L2 (Ramirez and Chiodo 1994). For example, the comma and the decimal point are used in Europe and some South American countries in the opposite way from the way they are used in the United States. The comma is used to separate whole numbers and decimals, and the period is used to separate integers. Students from those countries learned to write "1.223,04" instead of "1,223.04." In arithmetic computations, some Latin American students have been taught a placement of the divisor and dividend in an algorithm that is the opposite of that used in the United States.

Discourse Competence

To understand mathematical discourse, LEP students have to recognize the language functions of discourse. In the representative function, the mathematics discourse provides the discriptive information required to solve a problem. In the directive function, the mathematics discourse directs that the information be acted on (Kessler 1987). For example, in the word problem "Juan bought 20 marbles and gave Tom 5 and Mary 10. How many marbles does Juan have?" the first sentence serves the representative function, "He *bought* and he *gave*." The second sentence serves the directive function: "Determine how many he *has* now." Many native speakers of English could confuse these two functions. Only the verbs give clues to distinguish them. These difficulties are increased for LEP students because verbs in L2 are a source of confusion in distinguishing between descriptions and instructions or directions. In another example, LEP students were asked to write the false proposition "three is greater than seven." Many of them wrote "3 < 7." It can be assumed that they used their knowledge of inequalities instead of following the verbal instructions in English. They confused the instruction to write a false proposition (directive function) with an instruction to describe the way things should be mathematically (representative function).

Sociolinguistic Competence

In the language of mathematics, as in everyday English, the cultural context is directly linked with the meaning of the message (Secada 1993). Not having the cultural experience that goes with the linguistic expression leads to misunderstanding the communication (Pérez and Torres-Guzmán 1992). Specifically in word problems, the lack of cultural information about the meaning of the terms and the context in which they have been used results in miscommunication and frustration. For example, the following problem (Fendel 1987, p. 54) was given to several high school students from El Salvador:

> Bill started running around the track at 2 P.M. His average time was 4 minutes for each lap. He finished at 2:32 P.M. How many laps did he run?

They understood the concept of average, but they could not solve the problem because they did not know what "running laps on a track" meant. "Running laps on a track" was not in their personal experience. When the notion of running around a soccer field was substituted for that of running laps, four out of six students were able to solve the problem correctly. Many English-speaking students could face the same kind of difficulties, but for LEP students, the problem involves more than a lack of vocabulary; it is a cultural problem (Secada 1993).

Strategic Competence

Strategic competence in mathematics communication can be defined as the ability to decode a mathematical message. Second-language learners with some years of schooling have already developed this competence in their mother tongue to some degree. The capability to transfer it from L1 to L2 is related to the level of mastery of the other competences: the grammatical, the discourse, and the sociolinguistic competences. For example, to solve the problem "Freda bought several pads of paper for a total cost of $4.00. How many pads did she buy, and what was the cost per pad?" (Fendel 1987, p. 55), L2 learners must use all the communicative competences. If they (1) understand what is a "pad" (lexicon), (2) understand that "bought several for a total cost of x" is all the information provided (representative function of the discourse), (3) understand what a "pad of paper" is (sociolinguistic), and (4) understand the directive function of the discourse that requires them to figure out how much a pad costs, then they will be able to respond appropriately. They are required to use their communicative strategic competence to decode the meaning of the message. The response sought here is to realize that the data are incomplete and that in order to solve the problem, more information should be provided. Since many possibilities exist for missing information, problems with missing information can have several different results. Prior experiences in using this strategy in L1 will suggest several correct solutions. At this point, transference from L1 to L2 is at work. The example suggests that without the other competences, the strategies learned in L1 are useless.

Mathematical Knowledge and Behaviors

We can proceed to discuss the mathematical knowledge and behaviors associated with the communicative competences. Referring to our model (see fig. 27.1), we find that mathematical language proficiency is related to students' communicative competence in mathematics, their language proficiency in L1 and L2, and their ability to transfer mathematical language from one language to the other. As the model shows, these communicative variables are also linked to the mathematical knowledge and behaviors necessary to perform in mathematics. This section describes the four categories of mathematical knowledge and behaviors suggested by Schoenfeld (1985) that are necessary to perform in problem-solving situations. These categories are used to discuss LEP students' specific problems in thinking "about" mathematics in English. Since this issue is fundamental in the mathematics education of LEP students, several classroom recommendations are suggested in each category.

Mathematics education requirements for LEP students are not different from those for English-speaking students. However, mathematical knowledge and behaviors present a special kind of problem for LEP learners. Whereas monolingual students use their mother tongue to learn how to think "about" mathematics, LEP students must learn how to think mathematically in a second language. They have to learn mathematics facts, concepts, and algorithms and develop problem-solving strategies in the new language. They have to learn how to decide on the best strategy in a specific situation and learn to see themselves as able to perform successfully in mathematics in L1 and L2. The prior mathematical knowledge and behaviors necessary to learn mathematics have been classified by Schoenfeld (1985) in four categories: the cognitive resources, the heuristics, the control process, and the belief system. Since these categories are the foundations for mathematics learning, they have been integrated into the model and used to suggest specific classroom recommendations.

Cognitive Resources

This category includes all the mathematics facts, concepts, and algorithms in the students' background knowledge of mathematics. Depending on the age and the schooling of the LEP learners, cognitive resources have been developed at different degrees in the native language. The teacher has to provide all possible means to allow LEP students to demonstrate their mathematics cognitive resources. Manipulatives, pictures, diagrams, and body language can help these students demonstrate what they know about facts, concepts, and algorithms. Because the process is complicated, it is not uncommon for teachers to make false assumptions about students' cognitive resources. For example, if the teacher assumes that the LEP student does not know enough mathematics, he or she may give a low-level or boring assignment when the student actually has the necessary mathematics background to solve a complicated problem but has limited language proficiency. Conversely, the teacher can assume erroneously that LEP students do not understand the language of a problem when their deficiencies lie in the mathematics background necessary to solve it.

Classroom recommendation. Allow LEP students to rely on their mother tongue to make sense of the mathematics communication. According to the instructional strategies suggested by Cuevas (1991), LEP students should have the opportunity to talk about mathematics with one another in the mother tongue. The use of L1 can best be facilitated in a classroom structure in which LEP students learn cooperatively in small groups (McGroarty 1989). LEP students can help one another while working in small groups even if the teacher is not able to communicate with them. Cooperative learning provides the conditions for nonnative English speakers with different degrees of bilingualism to work in both languages.

Two arguments support the use of L1 in the mathematics education of LEP students. First, according to the *Curriculum and Evaluation Standards* (NCTM 1989), background knowledge is the basis of any learning process in mathematics. Since most LEP students' knowledge is in their mother tongue, they should be allowed to use their L1 whenever necessary. Second, mathematical communicative competences acquired in L1 can be transferred to L2. Performing in either language has proved to be an effective means to increase bilingualism, and evidence indicates that degrees of primary-school bilingual development correlate with higher-order mathematical skills (Bamford and Mizokawa 1992).

Classroom recommendation. Use techniques that link second language learning with mathematical communicative competence (lexicon and syntax). LEP students must be helped not only to transfer their previous cognitive resources from L1 to L2 but also to develop new resources by increasing their mathematical communicative competence in L2. To do so, teachers can (1) create activities that facilitate the recognition and decoding of symbols and processes, (2) provide practice in matching words, symbols, and mathematical expressions through flash cards, puzzles, and oral and written exercises, (3) use such semantic organizers as semantic maps and concept maps to relate vocabulary and mathematics concepts (Olivares 1993), (4) use such word- or paragraph-completion tasks as cloze exercises, and (5) write and keep a notebook with a bilingual picture dictionary of the mathematics terminology that can be used for future references.

Heuristics

Heuristics are identified as general strategies used in problem solving. A common strategy observed in LEP students is to rely on their knowledge of algorithms to perform computations. Without further references to English, they can frequently participate in the mathematics classroom by performing computations. This participation, however, is not only limited but in word problems can also be misleading. A typical situation of this nature is the use of key words to identify the computation necessary in a word problem. For example, Chinese LEP students were given the problem "Young has missed 6 classes. He has missed 2 fewer than Hung. How many has Hung missed?" Several students gave a wrong answer. When they were asked in Chinese for the strategy they used, they explained that they relied on the key word *fewer* and employed subtraction instead of interpreting the word in context. The lack of reading comprehension in English is the main difficulty for LEP students in the use of heuristics. To compensate for their low English proficiency, they need other clues to understand the problem.

Classroom recommendation. Promote, encourage, and use multiple sources and vary forms of the same message to increase communication to and from LEP students. To communicate effectively with ESL (English as a second language) learners, manipulatives, pictures, and kinesthetic tools are essential at all levels. These resources are appropriate not only for young children but also for adolescents and adults. Body language, pictures, and objects should be used to present the same problem. Relying solely on verbal communication limits the possibilities for understanding the message. Giving meaning to different representations of the same message exposes students to several forms of discourse competence. Maintaining the coherence and cohesiveness of the message in each of its representations will help the students to figure out the problem and the most effective strategy to solve it.

Control Process

The control process makes up the psychological construct referred to as *metacognition* (Lester, Garofalo, and Kroll 1989). The function of the control process is for learners to monitor and evaluate their own mathematical performance. Because it works as a regulator of cognition, the control process requires a strong base of prior knowledge. To monitor and regulate how they learn facts, concepts, and algorithms, L2 learners must be familiar with those cognitive resources. Therefore, to develop the control process, LEP students should rely first on their mother tongue and later transfer it to English. For example, they can verbalize the strategy used to solve a specific problem. By explaining their performance, they will become aware of their own learning strategies. Metacognitive awareness of learning will enable them to select the most effective strategy and discard the others.

Classroom recommendation. Develop students' metacognitive learning strategies in mathematics. Since there is evidence that LEP students who use metacognitive learning strategies show substantial gains in solving problems (Chamot and O'Malley 1993), employing metacognitive processes has the potential to improve the mathematics performance of these students. Metacognition in mathematics has proved more successful when students use a three-step comprehension monitor (Hoover and Nolan 1993): (1) Develop a plan. Read the question carefully to state exactly what is needed. LEP students should work in small groups to identify the goal and select a strategy to reach it. (2) Evaluate the plan. Here again, LEP students can work bilingually in small groups while constantly monitoring the chosen strategy and making sure they are moving toward the planned goal. (3) Describe the strategy and evaluate the process. At this step, LEP learners should be able to display metacognitive awareness not only by identifying the strategy used but also by describing how they dealt with the stumbling blocks (control process) encountered during the problem-solving process.

Belief System

According to Lester, Garofalo, and Kroll (1989), the belief system constitutes "the individual's subjective knowledge about self, mathematics problem solving, and the topics dealt with in problem statements" (p. 77). For the second-language learner, this category represents an important element of the communication process. Self-perception and the perception of mathematics are connected with performance in mathematics. Top LEP mathematics students can become frustrated and feel incompetent in the second language. For example, a Korean student complained that he could solve all the classroom mathematics that was presented to him in numbers and symbols, but he fails so often with word problems that he does not want to attempt any more until he has enough English skills. He has alienated himself from problem-solving activities related to words.

Classroom recommendation. Help LEP students develop their oral, reading, and writing skills in L2 by the following practices: (1) Constantly question them about the main idea and supporting details in different pieces of writing. Mathematics teachers, as well as all teachers, should realize that teaching reading skills to LEP students is their responsibility. (2) Design activities that require interpreting the content of a paragraph or an everyday interaction. The ability to think and interpret a word problem is not limited to computing the numerical values. LEP students need to learn to read English with meaning and not only by decoding. (3) Ask students to read and interpret graphs, pictures, and other print materials. Teachers should accept both literal and rough translations from L1 to L2. Free interpretations, invented spelling, and other mistakes in the use of L2 must be accepted because the focus is on the content of the message and not on the use of English. (4) Furnish oral practice in reading formulas and equations in L2 and ask LEP students to check their equivalents in L1 if known. All these approaches will help LEP students develop confidence in their ability to perform mathematically in L2.

Classroom recommendation. Develop students' ability to solve problems in L2 by exposing them to a language activity. For example, the students can create their own word problem or express a real-world problem in mathematical terms. Since the purpose of this activity is to develop communicative strategic competence, the focus should be on the content of the message (the problem) and not on the form of it. At the beginning of the activity, allow invented spelling, pictures, native language, and other means of communication. Wait until the problem is understood, then ask the students to deal with any difficulties with the English in that specific problem. Remember that using key words can be misleading and limit the students' ability to see beyond simple computation.

LEP students can also be asked to practice problem-solving strategies using any means available to communicate. For example, they can apply the classic four-step strategy:

1. Visualize the problem by drawing pictures and diagrams, using manipulatives, and dramatizing or representing it in any other way. Label each element of the problem. It is acceptable to use pictures or manipulatives at any grade level. They are being used not only to understand the problem but also to communicate mathematically.

2. Question a possible solution by discussing it with other students, using L1 or L2. Look for more than one possible solution.

3. Check the evidence that may suggest the solutions.

4. Explain the rationale for adopting the strategy used. Remind LEP students that the use of the key-word strategy can be misleading and provide an example if possible.

Concluding Comments

This paper presents a model (see fig. 27.1) that describes the learning conditions of LEP mathematics students and on the basis of that model, makes several recommendations for classroom practice. These recommendations emphasize the variables that should be considered when creating the learning environment needed by these students. It is up to teachers to create activities that will meet their students' specific needs. Those activities should expose LEP students not only to mathematical knowledge but more important, to mathematical language proficiency in English.

The relations among the variables presented in this model demonstrate that important differences exist between L2 learners and monolingual students in the mathematics classroom. These differences go beyond problems with the simple transference of concepts and terminology. They include mastering the components of mathematical language proficiency in L2. Because mathematical language proficiency affects the ability to communicate "about" mathematics, LEP students should be exposed to a learning environment that focuses on mathematical communication in the second language. LEP students can always rely on the computational skills they developed in L1, but until they achieve mathematical language proficiency in English, they will continue being mathematically limited in the new language.

References

Bamford, Kathryn W., and Donald T. Mizokawa. "Spanish Immersion Children in Washington State: Fourth Year of a Longitudinal Study." Paper presented at the annual meeting of the American Research Association, San Francisco, 20–24 April 1992. (ERIC Document Reproduction Service no. ED 350875)

Chamot, Anna U., and J. Michael O'Malley. *The CALLA Handbook: How to Implement the Cognitive Academic Language Learning Approach.* Reading, Mass.: Addison-Wesley Publishing Co., 1993.

Cuevas, Gilbert J. "Developing Communication Skills in Mathematics for Students with Limited English Proficiency." *Mathematics Teacher* 84 (March 1991): 186–89.

Fendel, Daniel M. *Understanding the Structure of Elementary School Mathematics.* Newton, Mass.: Allyn & Bacon, 1987.

Fox, Lynn H. "The Effects of Sex Role Socialization in Mathematics Participation and Achievement." In *Women and Mathematics: Research Perspectives for Changes,* edited by Lynn H. Fox, Elizabeth Fenemma, and J. Sherman, pp. 1–77. N.I.E. Papers in Education and Work, no. 8. Washington, D.C.: National Institute of Education, 1977.

Geva, Esther, and Ellen B. Ryan. "Linguistic and Cognitive Correlates of Academic Skills in First and Second Language." *Language Learning* 43 (March 1993): 5–42.

Hoover, Linda A., and James F. Nolan. "Reading in Math." In *Reading across the Curriculum: A Research Report for Teachers,* edited by Mary Dupnis and Linda

Merchant, pp. 64–76. Bloomington, Ind.: ERIC Clearinghouse on Reading and Communication, 1993.

Kessler, Carolyn. "Linking Mathematics and Second Language Teaching." Paper presented at the Twenty-first Annual Meeting of Teachers of English to Speakers of Other Languages, Miami Beach, Fla., 21–27 April 1987. (ERIC Document Reproduction Service no. ED 289357)

Lester, Frank K., Joe Garofalo, and Diana Lambdin Kroll. "Self Confidence, Interest, Belief, and Metacognition: Key Influences on Problem-Solving Behavior." In *Affect and Mathematical Problem-Solving: A New Perspective,* edited by Douglas B. McLeod and Verna M. Adams, pp. 75–87. New York: Springer-Verlag New York, 1989.

Mather, Jeanne Ramirez Corpus, and John J. Chiodo. "A Mathematical Problem: How We Teach Mathematics to LEP Elementary Students." *Journal of Educational Issue of Language Minority Students* 13(1994): 1–12.

McGroarty, Mary. "The Benefits of Cooperative Learning Arrangements in Second Language Instruction." *NABE Journal* 13 (Winter 1989): 127–43.

National Council of Teachers of Mathematics. *Curriculum and Evaluation Standards for School Mathematics.* Reston, Va.: National Council of Teachers of Mathematics, 1989.

Olivares, Rafael A. "The Language and Content Connection in the Education of Limited English Proficient Students." In *English Language Arts and the Children at Risk,* edited by Philip M. Anderson, pp. 85–97. New York: New York State English Council, 1993.

Pérez, Bertha, and María E. Torres-Guzmán. *Learning in Two Worlds: An Integrated Spanish/English Biliteracy Approach.* White Plains, N.Y.: Longman Publishing Group, 1992.

Santiago, Felicita, and George Spanos. "Meeting the NCTM Communication Standards for All Students." In *Reaching All Students with Mathematics,* edited by Gilbert Cuevas and Mark Driscoll, pp. 133–45. Reston, Va.: National Council of Teachers of Mathematics, 1993.

Schoenfeld, Alan H. *Mathematical Problem Solving.* Orlando, Fla.: Academic Press, 1985.

Secada, Walter G. *Teaching Mathematics with Understanding to Limited English Proficient Students.* Urban Diversity Series no. 101. New York: Columbia University, New York Institute for Urban and Minority Education, 1990. (ERIC Document Reproduction Service no. ED 322284)

———. "Toward a Consciously-Multicultural Mathematics Curriculum." In *Reinventing Urban Education: Multiculturalism and the Social Context of Schooling,* edited by Francisco L. Rivera-Batiz, pp. 235–55. New York: IUME Press, 1994.

Tamamaki, Kinko. "Language Dominance in Bilinguals' Arithmetic Operations according to Their Language Use." *Language Learning* 43 (June 1993): 239–62.

U.S. Department of Education, National Center for Educational Statistics. *The Condition of Education 1994.* Washington, D.C.: U.S. Department of Education, National Center for Educational Statistics, 1994.

28

Mathematics as a Language

Zalman Usiskin

Philosophy is written in this grand book, the universe, which stands continually open to our gaze. But the book cannot be understood unless one first learns to comprehend the language and read the letters in which it is composed. It is written in the language of mathematics, and its characters are triangles, circles, and other geometrical figures without which it is humanly impossible to understand a single word of it; without these, one wanders about in a dark labyrinth.

—Galileo, *The Assayer*

At the time of this famous remark (1623), algebra was in its infancy, and analytic geometry had yet to be invented. There was no probability theory. The basic vocabulary and symbols of calculus would be first introduced sixty years later by Newton and Leibniz. The $f(x)$ function notation, the symbol for π, and the abbreviations *sin* and *cos* would not come into the language until Euler popularized them more than a hundred years later. The inventions of statistical displays such as bar graphs and circle graphs were even farther in the future, and statistical theory was nonexistent.

The language Galileo wrote about was only a small piece of the edifice that today we call mathematics. Today's mathematical language underlies financial dealings worldwide, describes a wealth of characteristics of all sorts of phenomena, is integral in high-speed communication through words and pictures, and models far more of the physical world than could have been modeled with the mathematics known in Galileo's time.

The growing use of mathematical language throughout the world has only in recent times demonstrated Galileo's foresight. When most people say, "Mathematics is a language," they seem to be saying that mathematics has much in common with other languages. Thus they do not mean that mathematics is a language in the same sense that English is but that mathematics is *like* a language in many ways. For instance, one mathematics educator has called mathematics an "extension of language" (Weinzweig 1982). We often hear mathematics spoken of as a "formal language" or as a "symbolic language." Each of these designations serves to separate mathematics from other languages and to make one feel a priori that the learning of mathematics should be different from the learning of language. The purpose of this paper is to convince the reader that, despite these differences, mathematics is a language in the same sense that English or Japanese or French is a language, in the same sense that a parallelogram is a quadrilateral. Like most modern languages, mathematics is both oral and written and can be either informal or formal. Like all languages, communication is one of its major purposes. Like all languages, it not only describes concepts but helps shape them in the mind of the user. Like all languages, mathematics has its unique characteristics. That is, mathematics is *like* a language because it *is* a language like any other. (Those interested in this topic might also wish to read the paper by Schweiger [1994], which came to the attention of this author after this paper had been drafted.) It will follow then that just as the properties of all quadrilaterals apply to all parallelograms, we should look at how all languages are taught and learned and see what messages we can gain to help us guide the teaching and learning of mathematics.

CHARACTERISTICS OF LANGUAGE

Every language has a grammar, and a number of terms introduced in today's schoolbooks make it obvious that mathematics has a grammar like that of spoken languages: $3 + 4x$ and $56.2 - 1/5$ are called *expressions,* and $x = 2$ and $3x + y < 50$ are *sentences.* Some books call $=$, $\perp$, and $\cong$ *verbs.* Indeed, the language of mathematics has such a well-constructed syntax that it is used (in mathematical linguistics) to study other languages.

Is mathematics a language according to the characteristics of language used by language arts educators and students of linguistics? One textbook (Flood and Salus 1984, p. 4) states, "The essence of language is speech and the psychological realities underlying it. While writing is secondary, we are reduced to using writing systems to represent speech since sound is transitory and most scholarship is eye-oriented." Yet for us in mathematics, writing is not "secondary"; it is preeminent, and we would never say that we are "reduced" to using writing systems. We often prefer writing to speech because we feel writing conveys mathematical ideas more accurately and is less likely to be misinterpreted. It has nothing to do with "eye-orientation."

Yet the characterizations of language used by some theorists in linguistics could encompass mathematics. For Chomsky (1986, p. xxvi), the human "language faculty" is "the innate component of the mind/brain that yields knowledge of language when presented with linguistic experience, that converts experience to a system of knowledge." Because writing does "convert experience to a system of knowledge," Chomsky's characterization could apply as well to mathematics. His view that language gives us insights into the brain yields one reason why speech rather than writing is considered primordial, for one thinks of speech as having direct centers in the brain, whereas writing does not. Another reason obviously is that children learn to speak their native language before they learn to write.

The first chapters of an encyclopedia for students of language (Collinge 1990) reflect this order of formation. Language is considered in turn as available sound, as organized sound, as form and pattern, as a mental faculty, and only somewhat later as a written medium and as a spoken medium. In its preface, however, there is a more general characterization that surely includes mathematics: "Language is, after all, the medium of human interaction" (p. 15). Yet this encyclopedia contains no referents to mathematics, and a chapter entitled "Language and Computation," dealing with the impact of the computer in a wide range of linguistic areas, ignores computer languages completely!

Thus perhaps the reason that mathematics is not treated by scholars as a language in the same sense that English or Japanese or French is a language is that, except for the small whole numbers, mathematics does not tend to originate as a spoken language. Indeed, the promotion of vocal discourse in the mathematics classroom, as recommended in the *Professional Standards for Teaching Mathematics* (National Council of Teachers of Mathematics 1991), is relatively new. It used to be that we would encourage silence in the mathematics classroom and discourage communication among students. Recognizing that mathematics is a language forces one to rethink its teaching.

Of course, mathematics is a special language. But every language has its unique properties. For instance, English is unique in the influence that it has had on the culture of people not only on the British Isles and what were formally parts of the British empire but throughout the world. English did not have words for two types of lava from volcanos, so we borrowed *aa* and *pahoehoe* from Hawaiian. Eskimo languages have more words for *ice* than any other. There are more words for *camel* in Arabic than in any other language. Chinese is special for the number of its written characters. Xhosa and some other African languages are special for the clicks required in speaking them. Mathematics is special for the reasons Galileo gave, for the way it uses deduction, for its ability to solve a wide variety of problems. But as a language, I believe mathematics has more in common with other languages than it has differences. Another purpose of this paper is to point out that we have thought of mathematics as *too* special.

MATHEMATICS AS A WRITTEN LANGUAGE

It is often said that mathematics is a symbolic language. This is another way in which we are conditioned to view mathematics as something special. But there are no more symbols in elementary mathematics than there are in English, and far fewer symbols than in Chinese. The symbols of mathematics, like the letters or characters in other languages, form the written language of mathematics.

Languages borrow symbols from one another. Most of the letters we use to write English came from Latin. The Γ in the Cyrillic alphabet comes from Greek. Similarly, algebra uses the Latin alphabet; geometry and trigonometry borrow π, ϕ, and θ from Greek; and set theory employs $\aleph$ from Hebrew.

Languages also borrow words from one another. Our culture's mathematics has borrowed from many languages: *ellipse, parabola,* and *hyperbola* from Greek; *algorithm* and *algebra* from Arabic; *circle* and *radius* from Latin; and *slope* from native English (for many other examples, see Schwartzman [1994]). What is significant is that the borrowing has also gone the other way; words that have their original meanings in mathematics have come into the English language. For instance, the word *triangle* began as a mathematical term, but it has developed English meanings to describe a musical instrument or a relationship involving three persons.

Sometimes we don't know whether we are in one language or another. When we talk about the dish called a *soufflé,* we do not know whether we are speaking in French or English. Similarly, mathematics has so permeated our native languages (e.g., English) that we don't know and can't tell where our language stops and mathematics begins. Consider the following English sentences. Which contain numbers and which do not?

1. Sylvia has three brothers.
2. Bob is first in line.
3. My telephone number is 1–800–555–1212.
4. I live at 1234 Bolyai Boulevard.
5. That pizza costs $10.95.
6. Half of the senators voted for the bill.
7. In the 1980s there was an increase in the percent of students taking four years of science in high school.

It is my opinion that each of these sentences contains at least one number (in order: three, first, 18005551212, 1234, 10.95, half, four). But when I have asked teachers to look for numbers on pages of newspapers, in every group there are some who will question whether these instances of written language represent numbers. This view seems to arise from such a strong belief in the "specialness" of mathematics that when mathematics appears in a context in which it is so familiar (e.g., together with English in a newspaper article) that it is understood by everyone, then the mathematics is thought not to be mathematics.

By ignoring the most familiar mathematics, we make it seem as if mathematics is less common than it is, and so we make it seem more foreign a language than it is. This affects students' learning. Consider sentence 5 above. Though the *10.95* in $10.95 is a decimal, to many students and teachers it is money, not mathematics. Nor is the unit dollars treated like other units; in most schoolbooks, unlike meters or feet, monetary units are not considered as units of measure. It may be for this reason that some students can add monetary amounts but cannot perform the same additions with decimals. Or consider sentence 4 above.

One's street address number is clearly a number, and often its being odd or even gives it certain properties, and its size determines its location. Yet many have said that very young children cannot deal with large numbers, even though children go to school knowing many large numbers like addresses.

Sentence 7 contains quite a bit of mathematics: *1980s* signifies the interval from 1980 to 1989. The word *increase* is a mathematical term signifying a subtraction of some kind that resulted in a positive answer. The word *percent* is a synonym for *fraction,* itself signifying that the operation of division is lurking behind the thought. The quantity "four years" can be thought of as the number "four" followed by the unit "year." One could even argue that the term *high school* conveys some mathematical information because here it is a synonym for *grades 9 through 12.* Despite all the mathematics in this sentence, however, it is likely that most students would not view this sentence as being particularly mathematical, wrongly viewing only *numerals* as conveying mathematics. Even many educators would not acknowledge the language of mathematics in sentence 7; they might call it "technical English" rather than "untechnical mathematics."

Words such as *increase* and *rate* and phrases such as *exponential growth* and *circular reasoning* represent mathematical language that has come to pervade the English language. Some other words have meanings in English that are different from their mathematical meanings. For example, *if* in mathematics almost never means *if and only if,* as it often does in everyday use. *Random* has a precise meaning in probability that is definitely not *haphazard,* which is its everyday sense. But this observation is in some sense no different from noting that the word *elf* in German refers to the number 11 and does not refer to a small, often mischievous fairy. The same words often have different meanings in different languages.

Sometimes algebra is viewed as a language of its own, different from arithmetic. This is a reasonable view; one could identify algebra and arithmetic as two sublanguages of mathematics, in the same way that the set of rectangles is a subset of the set of parallelograms. But even then, the language of algebra is not special. As a written language, both algebra and arithmetic use many of the same symbols. But we are trained so much to think of algebra as more special than arithmetic that many people say that $3 + ? = 10$ is arithmetic but $3 + x = 10$ is algebra—and that $3 + \square = 10$ is somewhere in between. But in theory, these are simply three different ways of writing the same sentence; only our attitudes toward the question mark, the letter, and the box make them different.

MATHEMATICS AS AN ORAL LANGUAGE

It is well known that the most difficult number names for children in the United States to learn are those from eleven to nineteen. The reason is that they do not follow a pattern, as the 20s or 30s or any other decade does. In Chinese and Japanese, however, the spoken word for 11 is "10, 1," much like our spoken word for 21 is "20, 1." This pattern makes the number names easy for Chinese and Japanese children to learn. The oral language of mathematics, then, affects our ability to learn mathematics (e.g., see Miura et al. [1993]).

If a student does not know how to read mathematics out loud, it is difficult to register the mathematics because the oral language is essential for memory. An algebra student who cannot read $3x + 5 = 10$ as "three ex plus five equals ten" will have great difficulty internalizing what is meant when the teacher reads the sentence. I recall a corresponding difficulty when I was an undergraduate in an advanced mathematics course. The instructor wrote Ξ for the name of a function on the blackboard but did not speak its name (xi). It was difficult for me to think about this function without having a pronounceable name for it.

The spoken language of mathematics is important for the understanding of mathematical concepts. A cousin of mine was born deaf. When he was in high school, I was already teaching mathematics. Harboring the view that those who are hearing impaired would be more sensitive to visual stimuli, I asked him whether geometry or algebra is more difficult for deaf students to learn. He surprised me by answering that geometry is more difficult. I asked why, and he answered that geometry requires more explanation—and therefore a greater knowledge of language—than algebra does. His conclusion has been confirmed to me by teachers of hearing-impaired students.

My cousin's explanation is understandable when we realize that every modern language has its oral and written manifestations. Oral communication is necessary to understand mathematics, just as it is necessary to understand virtually every other language. Its importance becomes heightened when the mathematics is not algorithmic—when interpretations are required.

Freudenthal (1983) has noted that a mathematical definition constricts our view of ideas. Consider that in geometry, we normally define *angle* as "the union of two rays with the same endpoint." This is a static definition. It is mathematically very useful, for it establishes an angle as a set of points and it makes for relatively easy definitions of adjacent angles and vertical angles. But every teacher knows that the definition does not help students to understand angle measure. Somehow the teacher must convey the notion that some angles are "bigger" than others, even though they obviously have the same number of points. So we convey, with language and gestures, that when we are thinking of the measure of an angle, we are thinking of the amount of opening between the rays, of the difference in direction of the two rays. We immediately pass from the definition of angle to the concept of angle measure, a passage that requires oral communication.

When the definition of angle is changed in trigonometry so that it includes an initial side and a terminal side of an angle and allows angle measures to be any real numbers, the teacher cannot be content with just giving a new definition. There needs to be a discussion of the rationale for a change in definition, that is, the inadequacy of the geometry definition and the advantage of the trigonometry definition.

This example is one of many that could be given that demonstrates that the oral communication of mathematics is important for its learning. It enables the student to internalize the language and to connect it with other ideas. And it is necessary for the interpretation of the language.

MATHEMATICS AS A PICTORIAL LANGUAGE

Mathematics has a third form of expression, its pictorial or representational form. The most obvious examples of this form are in the coordinate graphs of functions and relations. For example, if we wish to communicate the notion that the area of a circle varies directly as the square of its radius (in its customary oral form), we can use the equation $A = kx^2$ (a written form) and from this, obtain its pictorial form, one-half of a parabola.

The pictorial language of mathematics contains more than coordinate graphs and is growing because of the ease with which computers enable the language to be drawn. It includes the trees and other networks of graph theory; the bar graphs, circle graphs, and stem-and-leaf diagrams of statistics; the Venn diagrams of logic; and other more exotic representations (e.g., see Sacco et al. [1987]). It is possible to add to this list all the concrete aids that are used in the teaching of mathematics—Cuisenaire rods, geoboards, Dienes blocks, algebra tiles, and the like.

Some might wish to consider pictures of squares and cylinders and other geometric drawings as part of the pictorial language. This question is more complex than it might seem. Some pictures are just that—pictures—and they do not constitute a different form of the mathematical language any more than pictures of objects constitute a different form of English. But since mathematics studies geometry, it also studies the properties of pictures, and at times the pictures themselves do constitute part of the language of mathematics. At these times, the creation of accurate diagrams and drawings, as well as the analysis of drawings, is itself part of mathematics.

As in any other expression of language, one must practice working in the pictorial form to master it. Once mastered, the pictorial forms of mathematics can be powerful aids to understanding the language (Schneider and Saunders 1980). They provide a different way of communicating the language that can be highly appealing.

We might be tempted to think of the pictorial language, too, as something special about mathematics, but other languages have their pictorial forms. The written language of music is pictorial, with symbols higher on the staff generally conveying tones of higher frequency. The hieroglyphics of ancient Egypt constituted a sophisticated pictorial language. Likewise, today's Tchokwe of Mozambique have a complex pictorial language that represents stories (Gerdes 1988).

MATHEMATICS AS A FOREIGN LANGUAGE

If mathematics is a language like English, then why is it so much harder to learn? One reason is that mathematics is a foreign language for many students, for it is learned almost entirely in school and it is not spoken at home. Learning a foreign language is far more difficult than learning one's native language. Native Chinese and Japanese have as much difficulty learning English as a second language as native English speakers have learning Japanese or Chinese, so the difficulty is not due to the particular language but something else.

What is a source of this difficulty? We know that native Japanese speakers have difficulty hearing the difference between the *l* and *r* sounds and often pronounce *r* as *l*. Similarly, native English speakers have great difficulty hearing the differences among the four tones in Chinese words. We think of the differences as subtle, but to the Chinese they are as great as the differences between our pronunciation of the word *project* as a noun ('proj-ect) and its pronunciation as a verb (pro-ject'). This example offers a significant lesson for the learning of oral language: If the oral language is not learned before a certain age, then physical limitations may develop that make it difficult or even impossible to learn the language.

This point has exceedingly important implications for the psychology of learning mathematics. It suggests that we may be missing students because we do not teach concepts early enough. It may explain why all students in some countries are considered able to encounter algebra and geometry in grades 7 and 8, whereas historically not all students in the United States and Canada have been thought to have this ability. Our delay in teaching these subjects may be creating limitations in our students that make it more difficult for our students to learn them.

MATHEMATICS AS A DEAD LANGUAGE

No one calls mathematics a dead language; it is more alive today throughout the world than it has ever been. But one can argue that the mathematics that is taught in some schools is treated more like Latin than like a living language. When a person who has studied any part of mathematics asks, "What was it for?" we know that we have taught that mathematics as if it were dead.

All languages change and evolve. One has merely to read any Old English poetry to see that words once in common use become obsolete. Mathematics, like all other languages, shares this property. The square-root algorithm, the use of logarithms for computation, and many more of our paper-and-pencil algorithms are obsolete. In the world at large, people use calculators to do all but the simplest arithmetic. Long division is dead. Long (partial-product) multiplication is dying. Computers and graphing calculators handle graphs. But in many classrooms and on many tests—both teacher made and standardized—students are still required to focus on these obsolete skills. If we teach mathematics without reference to its applications—if we teach mathematics that is no longer used outside of school—we are teaching mathematics as if it is a dead language. Dead languages are even harder to learn than living foreign languages because they lack currency.

MATHEMATICS AS A NONSENSE LANGUAGE

A greater amount of mathematics than we may think is taught as nonsense. When children are taught to memorize their multiplication tables before they have any knowledge of the various meanings and uses of multiplication, they are being asked to learn what is not nonsense to us but what is essentially nonsense to them. Many teachers think that because a child needs to be able to respond to a multiplication situation at random, the child should be taught that way. But just as words develop meaning by their relationships to one another, so do numbers and operations. To the child who knows that 7×9 is 63 but cannot relate 7×9 to 6×9 or to a physical situation, the multiplication facts are nonsense. The algebra student who can multiply $(3 + 4x)(2 + x)$ but has no way to check the answer is in the same position.

Psychologists sometimes purposely do experiments with nonsense syllables or words that have no relation to one another to test the ability of the human mind to learn and remember. These experiments tell us that the human mind is able to learn nonsense. It is capable of remembering words to which no meaning is attached. How many children learn the national anthem and have no idea what such words as *ramparts* and *perilous* mean? These examples are only a few of many that show that we are able to learn and remember things that have no meaning to us. But a nonsense language is surely more difficult to learn than even a dead language, for nonsense has neither context nor currency.

MATHEMATICS AS AN ABSTRACT LANGUAGE

Some have justified the learning of mathematics without context by evoking the view that mathematics is essentially an abstract language. This view is not without foundation. In the early part of the century, the great mathematician David Hilbert put forth the notion that mathematics is merely symbols that satisfy rules like those of an arbitrary game. Hilbert's approach was a natural outgrowth of his time, since questions had been raised about the logical consistency of mathematics. Hilbert's formalistic view of mathematics, as it was called, influences educators even today.

Another sense in which mathematics is abstract is that many mathematical ideas encompass the commonalities present in many types of situations without being encumbered by the differing details of these situations. So some have argued that it is easier to learn the one abstract notion first than to "abstract" the mathematics from the concrete (see, e.g., Adda [1982]).

From other languages, we know that abstraction per se does not necessarily cause difficulty. Ideas such as "honesty," "redness," and "force" are abstract, but children can learn aspects of them at very early ages. The problem is that often we have taught abstractions in mathematics without teaching the specifics that have led to the abstraction. One would never teach honesty without giving examples and nonexamples that mean something to the student; the same goes for redness and force. Whether one is teaching addition of fractions at the elementary school level, theorems about triangles at the secondary school level, or group theory at the college level, if the student is taught generalizations without rationales for the ideas, without specific examples, or without ties to what the student already knows, the learning will be more difficult.

MATHEMATICS AS A NATIVE LANGUAGE

The argument so far is that mathematics is a language like other languages, but it has been wrongly treated as if it is special. Furthermore, it is often taught as if it were a dead language or a nonsense language or a totally abstract language. It is for many people a foreign language, but we have not applied what we have learned about the learning of foreign languages to it. These views of the language of mathematics, and the educational policies and practices that arise from them, make it more difficult to learn mathematics than it need be.

In contrast, native languages are learned well by many. (In this regard, it is interesting to note that many of the Africans brought to the United States as slaves in the seventeenth and eighteenth centuries could speak three or more languages, but their intelligence was evaluated by their knowledge of English.) Thus, if we could make mathematics more like a native language, virtually everyone could learn a significant amount of mathematics. The ten-year-old who knows algebra is unusual but could be no more unusual than the ten-year-old who knows French if the environment were suitable.

The child learns the language the way those around her or him deal with it. If we speak only baby talk to a child, the child will learn the language as if baby talk *is* the language. If we do not use full sentences, the child will not. This is not to say that we should speak to babies as we do with adults, because we know that a baby's vocabulary is limited, but it does suggest that if we want the child to learn the adult's mathematics, our goal from the start should be the adult's version of the language.

This reasoning suggests that we must strive to remove artificial barriers that keep negative numbers and fractions from being discussed in early grades, that put algebra on the shelf until middle school or later, and that avoid calculus in high school. Such practices only confirm the obvious—that without exposure, one does not learn the language. They do not apply the fact that with repeated exposure and immersion, languages can be learned quickly.

Of course, everyone learns a language slightly differently from everyone else. Native English speakers have different vocabularies and use colloquialisms that are sometimes distinctive enough to identify an individual. And we all use informal English expressions on some occasions that we would never imagine using on formal occasions. Mathematics, too, has both formal and informal expressions—what some have described as school mathematics and street mathematics (Nunes, Schliemann, and Carraher 1993). The implications for teaching are that we should expect students to use the language of mathematics in different ways, both formal and informal, and that we should not try to unnecessarily restrict the freedom of mathematical expression of students.

We can learn from the teaching of reading and writing English. We know now that students learn to write well when they are not restricted to the words they can spell correctly. Furthermore, relaxing the restrictions makes them more willing to write (Graves 1983; Caulkins 1983). The analogy is that students ought to be given opportunities to apply mathematics not restricted to the numbers with which they can compute without error. Students learn to read best when given stories. This fact gives us another reason for teaching mathematics in context. Students do not increase their ability to read unless given material that is slightly beyond their current reading level. This suggests that if we wish students to grow mathematically, we should offer experiences that extend and apply what students know.

Yet, if we succeed in making mathematics a native language, there will be a cost to pay. Mathematics is today the province of an intellectual elite, just as reading was some centuries ago. It was thought then that one had to be specially endowed to learn to read. Now, with the universality of reading, we know that reading is attainable by virtually everyone, and it has lost its place as an indicator of talent. When mathematics reaches this point, people will no longer believe that it takes a special intelligence to understand mathematics. People who major in mathematics will no longer be viewed as special. The benefit, of course, is much greater. As Galileo knew, mathematics provides a key to understanding our world and communicating in it. He could not foresee the extent to which his remarks were understated.

Concluding Comments

This paper takes the stance that mathematics is a language and that many of the difficulties children have in learning mathematics may be due to our treatment of mathematics as a dead language or a nonsense language, types of languages that are of course more difficult to learn than a living second language or one's native language. A generation ago we tended to delay teaching the living second languages until high school and taught the formal grammar before the student understood what the language is about. Now we have learned that conversation and immersion in situations are very effective ways to teach foreign languages and that virtually all young children can learn second languages when immersed in them. Because mathematics is a living second language for most students, at a minimum we should teach mathematics as we do living foreign languages—in context, starting as early as we can and immersing students in the language.

Of all languages, one's native language is the easiest to learn. Thus the increasing proliferation of mathematical language in the world at large is a boon to the learning of mathematics because it means that increasing amounts of mathematics are encountered by students in their lives outside of school and that more and more mathematics is becoming part of every student's native language repertoire. Furthermore, unlike native tongues, much of the written mathematical language is the same for a majority of the world's population. This expanding base for mathematics promises a citizenry of the future that better understands the world and can more easily communicate in it.

REFERENCES

Adda, Josette. "Some Aspects of the Relationship to Mathematics of Children Who Fail in Elementary Schooling." In *Language and Language Acquisition,* edited by Frances Lowenthal, Fernand J. Vandamme, and Jean Cordier, pp. 297–302, 334–40. New York: Plenum Press, 1982.

Caulkins, Lucy McCormick. *Lessons from a Child.* Portsmouth, N.H.: Heinemann Books, 1983.

Chomsky, Noam. *Knowledge of Language: Its Nature, Origin, and Use.* New York: Praeger Publishers, 1986.

Collinge, Neville Edgar, ed. *An Encyclopedia of Language.* London: Routledge, 1990.

Flood, James, and Peter H. Salus. *Language and the Language Arts.* Englewood Cliffs, N.J.: Prentice-Hall, 1984.

Freudenthal, Hans. *The Didactical Phenomenology of Mathematical Structures.* Dordrecht, Holland: D. Reidel, 1983.

Gerdes, Paulus. "On Possible Uses of Traditional Angolan Sand Drawings in the Mathematics Classroom." *Educational Studies in Mathematics* 19 (1988): 3–22.

Graves, Donald. *Writing: Teachers and Children at Work.* Portsmouth, N.H.: Heinemann Educational Books, 1983.

Miura, Irene T., Yukari Okamoto, Chungsoon C. Kim, Marcia Steere, and Michel Fayol. "First Graders' Cognitive Representation of Number and

Understanding of Place Value: Cross-National Comparisons—France, Japan, Korea, Sweden, and the United States." *Journal of Educational Psychology* 85 (1993): 24–30.

National Council of Teachers of Mathematics. *Professional Standards for Teaching Mathematics*. Reston, Va.: National Council of Teachers of Mathematics, 1991.

Nunes, Terezinha, Analucia Dias Schliemann, and David William Carraher. *Street Mathematics and School Mathematics*. London: Cambridge University Press, 1993.

Sacco, William, Wayne Copes, Clifford Sloyer, and Robert Stack. *Glyphs*. Dedham, Mass.: Janson Publications, 1987.

Schneider, Joel, and Kevin Saunders. "Pictorial Languages in Problem Solving." In *Problem Solving in School Mathematics,* 1980 Yearbook of the National Council of Teachers of Mathematics, edited by Stephen Krulik, pp. 61–69. Reston, Va.: National Council of Teachers of Mathematics, 1980.

Schwartzman, Steven. *The Words of Mathematics*. Washington, D.C.: Mathematical Association of America, 1994.

Schweiger, Fritz. "Mathematics Is a Language." In *Selected Lectures from the 7th International Congress on Mathematical Education,* edited by David F. Robitaille, David H. Wheeler, and Carolyn Kieran, pp. 297–309. Sainte-Foye, Quebec: Les Presses de l'Université Laval, 1994.

Weinzweig, Avrum I. "Mathematics as an Extension of Language." In *Language and Language Acquisition,* edited by Frances Lowenthal, Fernand J. Vandamme, and Jean Cordier. New York: Plenum Press, 1982.

Index